FORD

SHOP MANUAL

More information available at haynes.com
Phone: 805-498-6703

Haynes Group Limited
Haynes North America, Inc.

ISBN-10: 0-87288-753-7
ISBN-13: 978-0-87288-753-4

Disclaimer

There are risks associated with automotive repairs. The ability to make repairs depends on the individual's skill, experience and proper tools. Individuals should act with due care and acknowledge and assume the risk of performing automotive repairs.

The purpose of this manual is to provide comprehensive, useful and accessible automotive repair information, to help you get the best value from your vehicle. However, this manual is not a substitute for a professional certified technician or mechanic.

This repair manual is produced by a third party and is not associated with an individual vehicle manufacturer. If there is any doubt or discrepancy between this manual and the owner's manual or the factory service manual, please refer to the factory service manual or seek assistance from a professional certified technician or mechanic.

Even though we have prepared this manual with extreme care and every attempt is made to ensure that the information in this manual is correct, neither the publisher nor the author can accept responsibility for loss, damage or injury caused by any errors in, or omissions from, the information given.

Information and Instructions

This individual Shop Manual is one unit of a series on agricultural wheel-type tractors. Contained in it are the necessary specifications and the brief but terse procedural data needed by a mechanic when repairing a tractor on which he has had no previous actual experience.

The material is arranged in a systematic order beginning with an index which is followed immediately by a Table of Condensed Service Specifications. These specifications include dimensions, clearances, capacities and tune-up information. Next in order of arrangement is the procedures section.

In the procedures section, the order of presentation starts with the front axle system and steering and proceeds toward the rear axle. The last portion of the procedures section is devoted to the power take-off and power lift systems. Interspersed where needed in this sec-tion are additional tabular specifications pertaining to wear limits, torquing, etc.

How to use the index

Suppose you want to know the procedure for R&R (remove and reinstall) of the engine camshaft. Your first step is to look in the index under the main heading of "Engine" until you find the entry "Camshaft." Now read to the right. Under the column covering the tractor you are repairing, you will find a number which indicates the beginning paragraph pertaining to the camshaft. To locate this paragraph in the manual, turn the pages until the running index appearing on the top outside corner of each page contains the number you are seeking. In this paragraph, you will find the information concerning the removal of the camshaft.

Common spark plug conditions

NORMAL
Symptoms: Brown to grayish-tan color and slight electrode wear. Correct heat range for engine and operating conditions.
Recommendation: When new spark plugs are installed, replace with plugs of the same heat range.

WORN
Symptoms: Rounded electrodes with a small amount of deposits on the firing end. Normal color. Causes hard starting in damp or cold weather and poor fuel economy.
Recommendation: Plugs have been left in the engine too long. Replace with new plugs of the same heat range. Follow the recommended maintenance schedule.

TOO HOT
Symptoms: Blistered, white insulator, eroded electrode and absence of deposits. Results in shortened plug life.
Recommendation: Check for the correct plug heat range, over-advanced ignition timing, lean fuel mixture, intake manifold vacuum leaks, sticking valves and insufficient engine cooling.

CARBON DEPOSITS
Symptoms: Dry sooty deposits indicate a rich mixture or weak ignition. Causes misfiring, hard starting and hesitation.
Recommendation: Make sure the plug has the correct heat range. Check for a clogged air filter or problem in the fuel system or engine management system. Also check for ignition system problems.

PREIGNITION
Symptoms: Melted electrodes. Insulators are white, but may be dirty due to misfiring or flying debris in the combustion chamber. Can lead to engine damage.
Recommendation: Check for the correct plug heat range, over-advanced ignition timing, lean fuel mixture, insufficient engine cooling and lack of lubrication.

ASH DEPOSITS
Symptoms: Light brown deposits encrusted on the side or center electrodes or both. Derived from oil and/or fuel additives. Excessive amounts may mask the spark, causing misfiring and hesitation during acceleration.
Recommendation: If excessive deposits accumulate over a short time or low mileage, install new valve guide seals to prevent seepage of oil into the combustion chambers. Also try changing gasoline brands.

HIGH SPEED GLAZING
Symptoms: Insulator has yellowish, glazed appearance. Indicates that combustion chamber temperatures have risen suddenly during hard acceleration. Normal deposits melt to form a conductive coating. Causes misfiring at high speeds.
Recommendation: Install new plugs. Consider using a colder plug if driving habits warrant.

OIL DEPOSITS
Symptoms: Oily coating caused by poor oil control. Oil is leaking past worn valve guides or piston rings into the combustion chamber. Causes hard starting, misfiring and hesitation.
Recommendation: Correct the mechanical condition with necessary repairs and install new plugs.

DETONATION
Symptoms: Insulators may be cracked or chipped. Improper gap setting techniques can also result in a fractured insulator tip. Can lead to piston damage.
Recommendation: Make sure the fuel anti-knock values meet engine requirements. Use care when setting the gaps on new plugs. Avoid lugging the engine.

GAP BRIDGING
Symptoms: Combustion deposits lodge between the electrodes. Heavy deposits accumulate and bridge the electrode gap. The plug ceases to fire, resulting in a dead cylinder.
Recommendation: Locate the faulty plug and remove the deposits from between the electrodes.

MECHANICAL DAMAGE
Symptoms: May be caused by a foreign object in the combustion chamber or the piston striking an incorrect reach (too long) plug. Causes a dead cylinder and could result in piston damage.
Recommendation: Repair the mechanical damage. Remove the foreign object from the engine and/or install the correct reach plug.

SHOP MANUAL
FORD

MODELS 9N (Ford-Ferguson), 2N, 8N

The 9N, 2N and 8N model number designation depicts the year of introduction. The 9N was introduced in 1939 and was the earliest version. The 2N was introduced in 1942 and was the wartime version, with plans made to equip them with steel wheels, a magneto and other changes. Although some early 2N models were produced with steel wheels and no electrical system, most of the changes from the 9N never got into production. The 8N was introduced in 1948 and was in production through 1952.

The original tractor paint colors were as follows: Models 9N and 2N were painted a solid gray. The Model 8N hood, fenders and wheels were painted a lighter color of gray than the 9N and 2N gray; the front axle, engine, chassis, instrument panel and "Ford" script on the hood and fenders were painted red; and the radiator cap and seat were black. The gray paint for the 9N and 2N tractors and the gray and red paint for 8N tractors is available from Ford New Holland tractor dealers.

The tractor serial number is stamped on the left side of the engine block on 9N, 2N and 8N tractors. These Ford tractors do not have separate engine and tractor serial numbers.

TRACTOR SERIAL NUMBERS AND YEARS PRODUCED					
Models 9N-9NAN		Models 2N-2NAN		Models 8N-8NAN	
Year	Beginning Serial Number	Year	Beginning Serial Number	Year	Beginning Serial Number
1939	1	1942	99003	1947	1
1940	10234	1943	105375	1948	37908
1941	45976	1944	126538	1949	141370
1942	88888	1945	169982	1950	245637
1943	105412	1946	198731	1951	363593
		1947	258504	1952	442035

INTRODUCTION

This service manual covers all American made FORD Tractors manufactured from 1939 to 1952. This manual can be used by anyone with minimum experience and mechanical ability. Easy to read type, detailed drawings and clear photographs guide you through jobs ranging from simple maintenance to complete overhaul.

Where repairs are practical for the owner/mechanic, complete procedures are given. Where special tools are required and recommended, their designations are provided. Such tools may often be borrowed or rented, or can be purchased from a local Ford New Holland dealer or directly from a tool company such as the Owatonna Tool Co., 436 Eisenhower Dr., Owatonna, Minnesota 55060.

INDEX (By Starting Paragraph)

DUAL DIMENSIONS

This service manual provides specifications in both U.S. Customary and Metric (SI) systems of measurement. The first specification is given in the measuring system perceived by us to be the preferred system when servicing a particular component, while the second specification (given in parenthesis) is the converted measurement. For instance, a specification of 0.011 inch (0.28 mm) would indicate that we feel the preferred measurement in this instance is the U.S. Customary system of measurement and the Metric equivalent of 0.011 inch is 0.28 mm.

CONDENSED SERVICE DATA

Models
9N, 2N, 8N

GENERAL

Engine Make	Own
Engine Type	L-Head
Number of Cylinders	4
Bore	3.187 in. (80.96 mm)
Stroke	3.750 in. (95.25 mm)
Displacement	119.7 cu. in. (1.9 L)

Power Rating at Belt Pulley—Maximum

9N, 2N	23.6 hp (17.6 kW)
8N	27.3 hp (20.4 kW)

Power Rating at Drawbar—Maximum

9N, 2N	16.3 hp (12.2 kW)
8N	23.2 hp (17.3 kW)

Compression Ratio—Gasoline:

9N, 2N, Early 8N	6:1
Later 8N	6.5:1
Pistons Removed From	Above
Main Bearings, Number of	3
Main Bearings, Adjustable?	No
Rod Bearings, Adjustable?	No
Cylinder Sleeves, Dry, Wet?	Dry

Production Cylinder Sleeves--

Material (8N Prior to S.N. 433578, 9N, 2N)	Steel
Material (8N After S.N. 433577)	Cast Iron

Service Cylinder Sleeves--

Material (All Models)	Cast Iron
Generator Make	Own
Maximum Output	20 Amps
Starter Make	Own
Type	6-Volt

Battery

Type	6-Volt
Ground Terminal	Positive

Models
9N, 2N, 8N

Tire Size—Standard	
Front	4-19 4-ply

Rear:

9N (Early)	8-32 4-ply
9N (Late)-2N-8N	10-28 4-ply

Transmission

Type	Constant Mesh
Forward Speeds (9N, 2N)	3
Forward Speeds (8N)	4

Hydraulic Pump

Type	Scotch Yoke Piston
Capacity @ 2000 Engine Rpm	2.85 Gal./min. (10.8 L/min.)

TUNE-UP

Firing Order	1-2-4-3
Valve Tappet Gap (Cold)	
Inlet	0.010-0.012 in. (0.26-0.30 mm)
Exhaust	0.014-0.016 in. (0.36-0.40 mm)

Valve Face Angle

Inlet and Exhaust	45°

Valve Seat Angle

Inlet and Exhaust	45°
Ignition Distributor Make	Own

Distributor Model

8N Prior to S.N. 263844, 9N, 2N)	9N12100
8N After S.N. 263843	8N12127

Breaker Point Gap

Angle Mounted Distributor

No. 8N12127	0.025 in. (0.63 mm)

Face Mounted Distributor

No. 9N12000	0.015 in. (0.38 mm)

Retarded Timing

8N Prior to S.N. 263844, 9N, 2N	TDC
8N After S.N. 263843	4° BTDC

Advanced Timing

8N Prior to S.N. 263844, 9N, 2N	25° BTDC

Advanced Timing

8N After S.N. 263843	17° BTDC

TUNE-UP (Cont.)

Flywheel Timing Mark Indicating:
Retarded Timing (8N Prior to
S.N. 263844, 9N, 2N)............... None
Retarded Timing (8N After
S.N. 263843)...................... 4° Line
Advanced Timing (8N Prior to
S.N. 263844, 9N, 2N)................ None
Advanced Timing (8N
After S.N. 263843)............... 17° Line
Distributor Governor Advance Curve... See Text
Spark Plug
Make........................ Champion
Plug Model for Gasoline.............. H10
Electrode Gap.............. 0.025-0.028 in.
(0.64-0.71 mm)

Carburetor
Make.................... Marvel-Schebler
Carburetor Model................. See Text
Carburetor Float Setting............. 9/32 in.
(7 mm)

Carburetor Initial Adjustment
Idle Adjustment Needle........ 1 Turn Open
Main Jet Adjustment Needle..... 1 Turn Open
Engine Low Idle RPM 400
Engine High Idle RPM 2200
Belt Pulley RPM @ 2000 Engine RPM 1358
Pto RPM @ 1500 Engine RPM 545
Compression Pressure @ Cranking Speed
Minimum .... 90 psi
(620 kPa)

SIZES—CLEARANCES

Crankshaft Journal Diameter.... 2.248-2.249 in.
(57.10-57.12 mm)
Crankpin Diameter .... 2.094 in.
(53.18 mm)
Camshaft Journal Diameter .... 1.797 in.
(45.64 mm)
Piston Pin Diameter .... 0.7501-0.7504 in.
(19.05-19.06 mm)
Valve Stem Diameter
One-Piece Valve Guide........ 0.341-0.342 in.
(8.66 mm)

Two-Piece Valve Guide....... 0.3105-0.3115 in.
(7.89-7.91 mm)
Cam Follower (Push Rod) Diameter ... 0.9995 in.
(25.38 mm)
Compression Ring Width............. 0.093 in.
(2.36 mm)
Oil Ring Width .... 0.187 in.
(4.75 mm)
Main Bearings Running
Clearance .... 0.001-0.003 in.
(0.025-0.076 mm)
Rod Bearings Running
Clearance .... 0.001-0.0035 in.
(0.025-0.089 mm)
Piston Skirt Clearance
Steel Pistons.............. 0.0025-0.004 in.
(0.064-0.101 mm)
Aluminum Pistons .... 0.0015-0.0025 in.
(0.038-0.063 mm)
Camshaft Bearing Clearance .... 0.001-0.002 in.
(0.025-0.050 mm)
Cam Follower (Push Rod)
Running Clearance.......... 0.0004-0.001 in.
(0.010-0.025 mm)
Crankshaft End Play .... 0.002-0.006 in.
(0.05-0.15 mm)

CAPACITIES

Cooling System.................... 3 Gallons
(11.3 L)
Crankcase Oil (With Filter Change).... 6 Quarts
(5.6 L)
Fuel Tank
Standard..................... 9 Gallons
(34 L)
Reserve .... 1 Gallon
(3.8 L)
Transmission, Differential &
Hydraulic System................ 5 Gallons
(18.9 L)
Belt Pulley Housing................. 1/3 Quart
(0.3 L)

*Fig. 1—MODELS
9N AND 2N*

Fig. 2—MODEL 8N

LUBRICATION AND MAINTENANCE

SCHEDULED MAINTENANCE

1. Scheduled maintenance tasks and checks should be performed at certain time or hourly intervals as outlined below. The item numbers in parenthesis refer to Fig. FO1.

Daily or Every 10 Hours of Operation
- Lubricate steering drag links (2 and 19) with grease.
- Lubricate steering spindles (4) with grease.
- Clean the air cleaner oil cup (6) and refill with engine oil. (May require attention more often under extreme conditions.)
- Lubricate the clutch linkage (7) with grease.
- Clean the oil filler tube breather cap (12) with solvent.
- Lubricate distributor oil cup (14) with a few drops of oil.
- Check engine coolant level.
- Check engine oil level on dipstick (16).
- Check hydraulic system oil level (21).
- Lubricate hitch lift arms (23) with grease.
- Check fuel tank sediment bowl and drain water and sediment if necessary.

Weekly or Every 50 Hours of Operation
- Check belt pulley (24) oil level and add SAE 90 EP gear lubricant if necessary.
- Check battery electrolyte level.
- Check tire air pressure.
- Check fan belt tension.

Every 100 Hours of Operation
- Change engine oil (25) and oil filter (5). See Note 1.

Every 200 Hours of Operation
- Lubricate generator rear bearing (1) with engine oil.

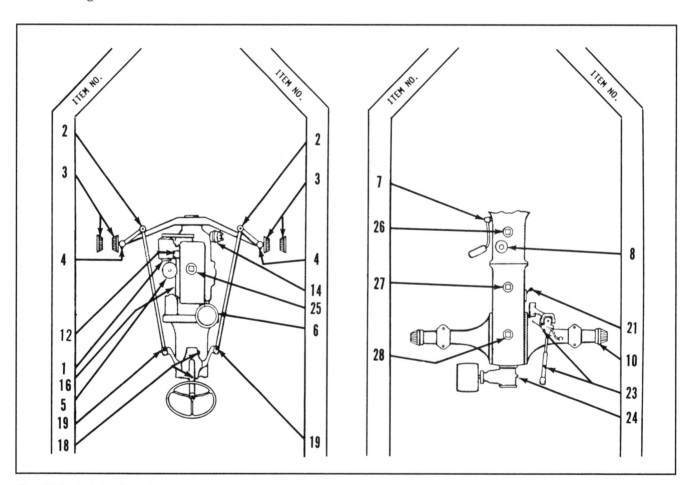

Fig. FO1—Lubrication chart.

1. Generator	6. Air cleaner	14. Distributor	23. Lift arms
2. Ball joint	7. Clutch pedal	16. Engine oil dipstick	25. Engine oil drain plug
3. Front wheel bearings	8. Oil filler plug	18. Steering gear	26. Drain plug
4. Spindle	10. Rear wheel bearing	19. Pitman arm	27. Drain plug
5. Oil filter	12. Crankcase breather	21. Transmission oil dipstick	28. Drain plug

- Clean distributor cam (14) and apply new lubricant to cam.
- Check steering gear (18) oil level and add SAE 90 EP gear lubricant if necessary. Fill to top of filler plug opening in side of steering gear housing.

Yearly or Every 600 Hours of Operation

- Tune-up the engine.
- Clean and repack front wheel bearings (3).
- Remove and clean carburetor air cleaner housing and filter element (6). See Note 2.
- Drain transmission, differential and hydraulic system oil and refill with new oil.
- Drain and flush cooling system. Refill with new coolant.

Every 1800 Hours of Operation

- Clean and repack rear wheel bearings (10).

NOTE 1 – If tractor is operated under any of the following conditions, change the engine oil and oil filter more frequently:
a. Extremely hot or cold temperatures
b. Sustained heavy loads
c. Extended low speed operation
d. Extremely dusty conditions

NOTE 2 – Under severe dust condition, remove and clean the carburetor air cleaner assembly every 100 hours of operation.

LUBRICATION

Engine Oil Change Periods

2. The frequency of oil changes depends upon the severity of operation. Under normal operation conditions, engine oil should be changed every 100 hours of operation. Under extreme conditions (dusty, high temperature and heavy loads), oil should be changed more frequently.

The oil should be changed every 50 hours when operating the tractor in below freezing temperatures. Intermittent engine operation and idling should be kept to a minimum in cold weather to avoid dilution of the oil. Low temperature operation promotes sludging which can plug oil passages and cause the formation of corrosive acids which result in rapid engine wear. Run the engine until the oil is at normal operating temperature prior to draining the oil. Remove the drain plug and allow the oil to drain for at least 10 minutes.

The oil filter element, located in filter canister on left side of the engine (Fig. FO2), should be changed at the same time as the oil. Always use new gaskets when reinstalling the filter canister and tighten retaining bolt to 20-25 ft.-lbs. (27-34 N•m) torque. Do not overtighten as canister may be distorted, resulting in oil leakage.

Crankcase oil capacity is 6 quarts (5.7 L) with a filter change. Select a good quality oil with SAE viscosity grade suitable for the ambient temperature. In the summer, consider the highest expected temperature. In the winter, the oil must be thin enough to permit easy starting.

Transmission, Differential and Hydraulic Oil

3. On 9N, 2N and 8N tractors, the transmission and differential housings serve as a common sump for the lubricating and hydraulic system fluid. The fluid level is checked by removing the level check plug located on the lower right side of the transmission housing on some early models, or by a dipstick located in the inspection plate on right side of differential housing on later models (Fig. FO3).

On all models, the lubricating and hydraulic fluid should be changed every 600 hours of operation or

Fig. FO2—View of left side of 8N engine.

Fig. FO3—A dipstick, located in inspection cover, is used to check transmission oil level.

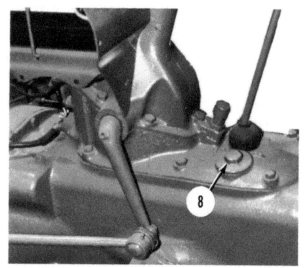

Fig. FO4—Transmission oil filler plug (8) is located on the transmission cover (9N shown).

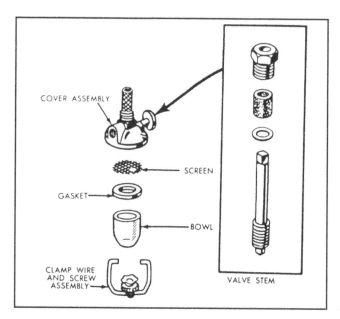

Fig. FO5—Exploded view of fuel shut-off valve and sediment bowl assembly.

once a year, whichever comes first. Refer to Fig. FO1 for location of drain plugs (26, 27 and 28) and to Fig. FO4 for location of filler plug (8). The fluid should be warm when it is drained, and all drain plugs should be removed on those models which have a common housing.

Fluid capacity is approximately 5 gallons (19 L) on 9N, 2N, and 8N tractors. The recommended fluid for use in all models is Ford M2C 134-D hydraulic fluid. It is also permissible to use extreme pressure gear lubricant in the transmission, differential and hydraulic system. Use SAE 90 EP gear oil if air temperature will be above 32° F (0° C) and SAE 80 EP gear oil if air temperature will be below 32° F (0° C).

MAINTENANCE PROCEDURES

Crankcase Ventilation

4. The engine crankcase is vented to the atmosphere to remove water vapor, gasoline vapor and blowby products which can cause deterioration of the oil and the corrosion of engine components. The crankcase is vented through the oil filler tube breather cap (Fig. FO2). If the crankcase ventilation system becomes restricted, the pressure in the crankcase will rise above normal. Higher than normal crankcase pressure may result in abnormal oil consumption and external oil leakage at the crankshaft seals.

The oil filler cap should be cleaned in suitable solvent after every 10 hours of operation.

Fuel System

5. To clean the fuel tank sediment bowl (Fig. FO5), turn fuel shut-off valve clockwise to closed position.

Loosen sediment bowl retaining nut, move retaining wire clamp sideways and remove the bowl. Clean the bowl with a clean cloth. Remove and clean the screen located in the bowl cover.

A strainer screen, attached to the top of sediment bowl cover, is located inside the fuel tank. In order to clean this screen it is first necessary to drain the fuel from the tank, then remove the cover assembly from the bottom of fuel tank.

A fuel strainer screen is also located in the fuel inlet elbow of the carburetor. The screen should be removed and cleaned periodically. The screen should also be cleaned if there is an unusually large accumulation of sediment in the fuel sediment bowl, or if there is an indication that not enough fuel is reaching the carburetor.

NOTE: If an excessive amount of dirt or water quickly accumulates in the sediment bowl, the fuel tank should be removed, drained and cleaned. Also, the source of the contamination must be found and corrected.

Carburetor Air Cleaner

6. An oil bath type air cleaner is used on all models (6—Fig. FO6). Air is drawn into the air cleaner and directed downward toward the oil in the cup attached to the bottom of the air cleaner body, then upward through the filtering element into the carburetor. Heavy particles of dirt are trapped by the oil in the cup, smaller particles not retained in the oil cup are screened out by the filter element in the air cleaner body. If the oil in the cup becomes thick and gritty, the air cleaner filter element will become restricted and cause a drop in the volume of air reaching the engine,

Fig. FO6—Air cleaner (6) is mounted on right side of tractor on all models. Air cleaner cup should be filled with clean engine oil to oil level mark (M).

resulting in loss of power and excessive fuel consumption.

The air intake on 9N and 2N tractors is beneath the hood and subjected to a sort of "dust trap" which has been corrected by most operators with the addition of an "Air Cleaner Extension" which raises the air intake above the hood. The air cleaner extension was available as a dealer option.

On the 8N tractor, a screened breather entrance is located on the outside of the right rear corner of the hood. A centrifugal type (Cyclone) attachment (31—Fig. FO6) was made available for 8N tractors which whirled the incoming air and collected the dry dust in a glass container where it could be emptied as the container became full.

The air cleaner cup should be removed and inspected every ten hours of operation, or more often under extremely dusty conditions. If oil in the cup is dirty, clean the cup and refill with fresh engine oil (same viscosity as used in engine crankcase) to the full mark (M—Fig. FO6) indicated on the cup.

Every 600 hours of operation, or more often under dusty conditions, remove the complete air cleaner from the tractor and wash the air cleaner body and filter element with a suitable solvent. Dry the filter element, then coat with light weight engine oil. Refill the oil cup to proper level with engine oil and reinstall the air cleaner assembly.

Failure to properly maintain the air cleaner will result in poor engine performance and premature wear of the engine.

Cooling System

7. A pressure-type cooling system is used on 2N and 8N models, which means that more heat is required to make the coolant boil than if the system were not under pressure. The radiator is sealed with a cap that contains a pressure valve and a vacuum valve.

The pressure valve in the radiator cap maintains cooling system pressure at 3.5 to 4.5 psi (24-31 kPa), raising the boiling point of the coolant many degrees. As the coolant heats, it expands, raising the pressure in the system. The vacuum valve in the cap admits displaced air when the engine is shut off, preventing damage to the system. When the system is operating properly, the air in the upper tank expands and contracts protecting the system from damage.

> **CAUTION: The radiator cap should never be removed when the engine is hot. However, if this is unavoidable, cover the cap with a thick cloth or wear heavy leather gloves. Slowly turn the cap counterclockwise against the first stop (about $1/4$ turn). Let all pressure (hot coolant and steam) escape, then depress the cap and turn counterclockwise to remove. If the cap is removed too soon, scalding coolant may escape and cause a serious burn.**

The coolant level should be checked at the beginning of the day when the engine is cold. Maintain the coolant level slightly below the bottom of the radiator filler neck to allow for expansion of the coolant when it reaches normal operating temperature.

Check the condition of the coolant. If it is dirty or rusty, drain the radiator and cylinder block and remove the thermostat. Clean and reverse flush the radiator and engine block, and refill with fresh coolant.

It is recommended that a mixture of clean water and ethylene glycol antifreeze be used as a coolant in summer as well as winter. Ethylene glycol antifreeze not only protects the coolant from freezing in cold weather, it also increases the coolant boiling point above that of plain water to reduce coolant loss during operation in hot weather. Ethylene glycol antifreeze also contains additives to inhibit the formation of corrosion and rust in the cooling system. Capacity of cooling system is 12 quarts (11.3 L) for 9N, 2N and 8N models.

If the temperature does not fall below the freezing point, plain water can be used in the cooling system; however, the cooling system should be treated with rust and corrosion inhibitor. If the water supply contains lime or alkali, it is recommended that distilled water or rain water be used for a coolant. Deposits caused by lime or alkali water will quickly build up on the engine coolant passages which will eventually result in overheating and possible engine damage.

Fan Belt Tension

8. The fan belt tension is adjusted by changing the position of the generator. See Fig. FO7. The belt tension is correct when the belt can be deflected ½ inch (13 mm) with moderate thumb pressure.

NOTE: The belt should be cool when tension is adjusted.

A new belt will stretch slightly during the first few hours of operation. When a new fan belt is installed, operate the engine for about two hours to "run in" the belt. Stop the engine then check and adjust the tension if necessary.

Do not overtighten the fan belt as premature wear of the generator and water pump bearings may result. If belt tension is too loose, belt slippage may occur, resulting in an overheated engine and low generator output.

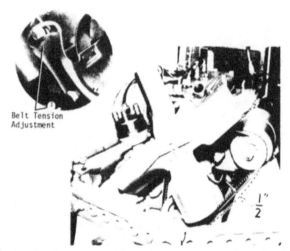

Fig. FO7—Change position of generator to adjust fan belt tension.

FRONT AXLE

TREAD WIDTH AND TOE-IN

9. ADJUSTMENT. The front axle is constructed in three sections and may be adjusted to vary the tread width from 48 inches (122 cm) to 76 inches (193 cm). See Figs. FO8 and FO9. Be sure to leave one or more open holes between the bolts attaching the axle extension to axle center member when changing the tread width. Note that the 76-inch (193 cm) tread width spacing is obtained by setting the axle for 68-inch (173 cm) tread and reversing the front wheels.

The rear wheel tread width settings can be adjusted to match the front tread width by changing the position of the steel discs and wheel rims. The rear tread width can be varied from 48 inches (122 cm) to 76 inches (193 cm) by installing wheel disks in either a convex or concave position and/or by installing rims on the disks in any of four different locations as shown in Fig. FO10.

NOTE: When making rear wheel tread width adjustments, be sure that arrow on the sidewall of the

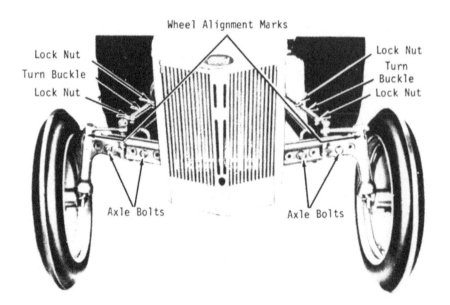

Fig. FO8—Front axle as used on 8N tractors. Note hole spacing of tread adjusting bolts.

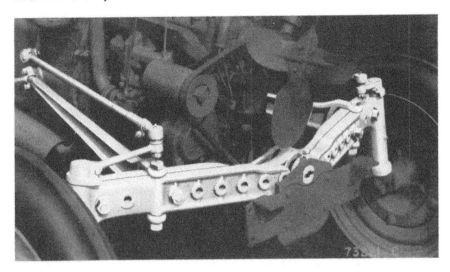

Fig. FO9—Front axle on 9N and 2N tractors.

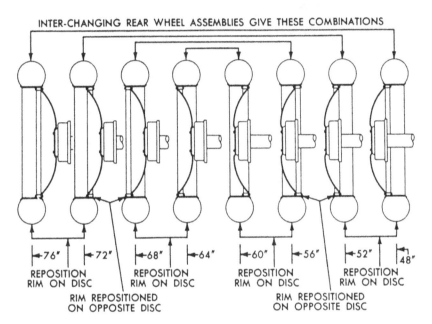

INTER-CHANGING REAR WHEEL ASSEMBLIES GIVE THESE COMBINATIONS

←76"→ ←72"→ ←68"→ ←64"→ ←60"→ ←56"→ ←52"→ ↑48"

REPOSITION RIM ON DISC   REPOSITION RIM ON DISC   REPOSITION RIM ON DISC   REPOSITION RIM ON DISC

RIM REPOSITIONED ON OPPOSITE DISC   RIM REPOSITIONED ON OPPOSITE DISC

Fig. FO10—Rear wheel tread width adjustment.

tire points in the direction of rotation of the wheel during forward travel.

Front wheel toe-in (front wheels closer together at the front than at the rear should be approximately ¼ inch (6.35 mm). When the toe-in was originally set at the factory, the spindle housing and steering arm were chisel marked as shown in Fig. FO8 to provide alignment marks for correct toe-in setting. If the alignment marks are not visible, measure distance between front of wheels and distance between rear of the wheels at hub height. Subtract front measurement from rear measurement to determine the toe-in.

Front wheel toe-in is adjusted by changing the length of the drag links. Be sure to adjust both drag links equally. On 9N and 2N models, the drag links must be disconnected from the steering arms. Loosen the lock clamps and thread end into or out of each drag link an equal amount as required. On 8N models, the drag links do not need to be disconnected to adjust the toe-in. Loosen the lock clamps on each of

the drag links, then turn the turnbuckles as required to obtain the desired toe-in.

FRONT WHEEL BEARINGS

10. REMOVE AND REINSTALL. The front wheel hubs and bearings (Fig. FO11) should be removed once a year or every 600 hours of operation for cleaning, inspection and lubrication.

To remove the front hubs and bearings, raise front of the tractor and remove the hub cap, cotter pin and wheel bearing adjusting nut on each wheel in turn. Pull the wheel (with the bearings) off the spindle. Drive the inner bearing and grease retainer out of the wheel hub. Clean all parts with a suitable solvent and inspect for wear or damage. Renew parts as necessary.

Pack the wheel bearings with grease prior to installing in wheel hub. Install inner bearing in hub,

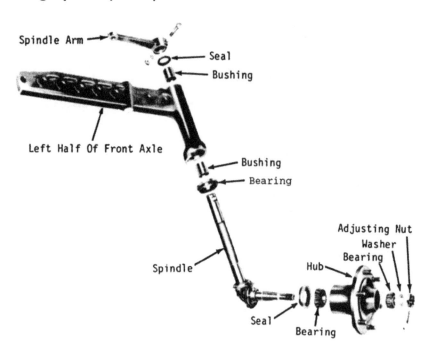

Fig. FO11—*Exploded view of spindle and wheel hub typical of all models.*

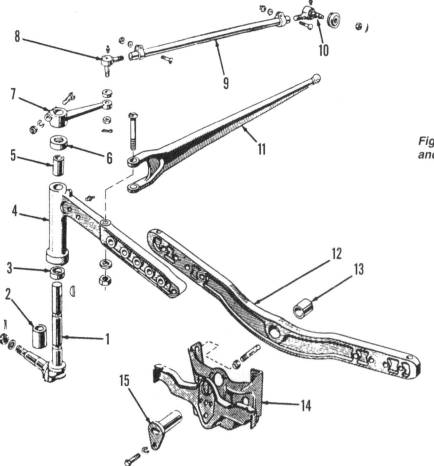

Fig. FO12—*Exploded view of front axle and related parts typical of all models.*

1. Spindle assy.
2. Lower bushing
3. Bearing
4. Axle extension
5. Upper bushing
6. Dust seal
7. Steering arm
8. Tie rod end
9. Tie rod sleeve
10. Tie rod end
11. Radius rod
12. Center axle
13. Pivot bushing
14. Front support
15. Pivot pin

then drive grease retainer into the hub. Place the hub, outer bearing and washer on the spindle. Install bearing adjusting nut and turn until tight, then loosen the nut approximately 1/8 turn. The wheel should turn freely, but there should be no visual looseness. Install cotter pin and hub cap.

SPINDLE BUSHINGS

11. REMOVE AND REINSTALL. To renew the spindle bushings (Fig. FO11), proceed as follows: Support the front of the tractor and remove the front wheels. Remove the steering arm and key from upper end of wheel spindle, then lower the spindle out of the axle extension. Drive the old bushings from the axle extension.

Install new bushings using a suitable piloted driver. New bushings are presized and will not require reaming if not distorted during installation. Renew the dust seals. Check and adjust front wheel toe-in if the spindles are renewed.

AXLE CENTER MEMBER AND KING PIN

12. REMOVE AND REINSTALL. To remove the axle king pin (15—Fig. FO12) and center member (12), raise and support the front of the tractor under the engine. Unbolt the radius rods (11) and axle extensions (4) from the axle center member. Swing the axle extensions and wheel assemblies away from the center member. Place a block of wood on top of the engine to support the hood and fuel tank. Remove the two cap screws attaching the hood to the axle front support and remove the grille. Remove the two cap screws securing the bottom of the radiator to the front axle support. Raise the radiator to provide clearance for removal of the king pin. Remove the cap screw attaching the axle king pin to the front support. Use hand crank as shown in Fig. FO13 or a suitable slide hammer to remove the king pin. Withdraw axle center member from either side of the front support.

The axle king pin bushing (13—Fig. FO12) can be renewed at this time. Make certain that the front axle king pin fits freely in bushing before reinstalling axle center member. Secure king pin locking flange with cap screw and lockwasher.

AXLE FRONT SUPPORT

13. REMOVE AND REINSTALL. To remove the axle front support (14—Fig. FO12), first raise front of tractor and support under the engine. Place a block on top of the engine to support the fuel tank. Remove the bolts attaching the front of the hood and bottom of the radiator to the front support. Remove the axle king pin (15). Remove the stud nuts or cap screws attaching the support to the engine. The support can now be lowered from the tractor.

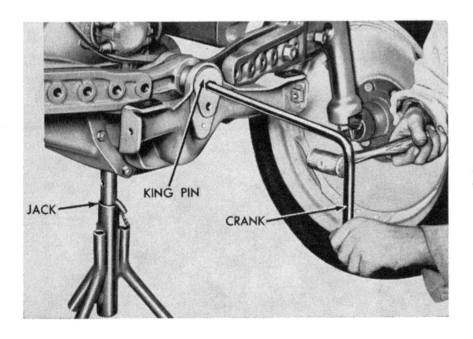

Fig. FO13—Method of removing front axle pivot pin.

STEERING SYSTEM

The steering system on Models 9N and 2N consists of two sector gears and integral arm units, both meshed with a steering shaft and pinion assembly. Gear units are mounted in a housing, which also serves as a transmission case cover. The gear housing is constructed in two parts and also supports the instrument panel. See Figs. FO14 and FO15.

STEERING GEAR

Models 9N-2N

14. ADJUSTMENT. The steering shaft bearings may be adjusted without removing steering gear unit from the tractor. To adjust steering shaft bearings, remove steering wheel and shaft bearing dust cover (19—Fig. FO16), spring (20) and seal. Remove upper locknut and lockwasher (1). Adjust lower nut until end play is removed and shaft still turns freely. Install lockwasher and locknut and tighten nut, then reinstall seal and dust cover. Detach drag links from sector arms and check backlash at sector arms while steering wheel is held stationary. If excessive sector backlash or end play exists, it will be necessary to remove upper gear housing (11) to gain access to sectors for adjustment.

To provide clearance for removal of steering gear upper housing, it will be necessary to first remove the hood and fuel tank, battery and instrument panel. Unbolt upper housing and invert it in a vise, then place both sectors (15 and 26) in position, meshed with steering shaft pinion (14). Check gear mesh and backlash with sectors held in normal operating position. Thrust washers (25 and 27), which are available in three thicknesses, should be installed as required for minimum sector end play and backlash. Reinstall the assembly after adjustment is complete and synchronize the sector gears (15 and 26) as follows.

When early Model 9N steering gear is turned to either extreme, severe steering conditions may cause sectors to jump out of mesh and result in an uneven steering radius. To synchronize the gears without removing the steering housing, disconnect both drag links at the steering arms and move left arm rearward as far as possible and right arm forward to unmesh the gears. Re-engage both sectors with steering shaft pinion and check synchronization by observing whether steering arms are parallel and point slightly rearward when gear is in mid or straight ahead position. Reconnect drag links and adjust as necessary.

15. R&R AND OVERHAUL. To overhaul the steering gear on 9N and 2N models, proceed as follows: Remove the battery, steering wheel, hood and instrument panel. Disconnect drag links from sector arms. Unbolt and separate the upper housing (11—Fig. FO16) from the lower housing (29). Remove nuts (1) from steering shaft (14) and withdraw shaft and pinion from upper housing. Renew worn parts and install new oil seals. Invert the upper housing in a

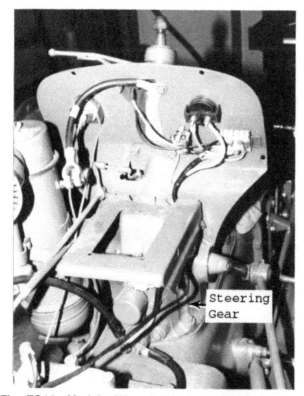

Fig. FO14—Models 9N and 2N steering gear unit with attached instrument panel and battery tray.

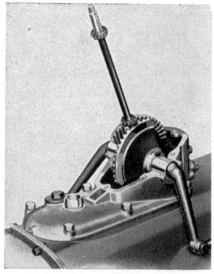

Fig. FO15—Model 9N and 2N steering gear with upper housing removed.

vise and place both sectors (15 and 26) in position, meshed with steering shaft pinion (14).

Check gear mesh and backlash with gears held in normal operating position. Thrust washers (25 and 27) are available in three thicknesses, and should be installed as required for minimum sector end play and backlash.

Reinstall assembly after adjustment is complete and synchronize the sector gears as follows: With the drag links disconnected, move the left steering arm rearward as far as possible, and right arm in opposite direction.

Re-engage both sectors with the steering shaft pinion and check synchronization by observing whether steering arms are parallel and pointing slightly rearward when gear is in mid or straight ahead position.

STEERING GEAR

Model 8N Prior to Serial Number 216989

The steering gear used on 8N tractors is a Saginaw screw and recirculating ball nut type, mounted on top of the transmission housing. On tractors prior to serial number 216989, the sector teeth are straight cut and the left sector (viewed from the seat on the tractor) meshes with the rack teeth on the ball nut. Refer to Fig. FO17. The right sector meshes with the left sector and rotates in the opposite direction.

16. ADJUSTMENT. Steering gear adjustment is correct when minimum backlash is present and steering performs properly. Sector shaft bearing housings on these tractors are slightly eccentric and are moved toward or away from ball nut and each other by rotating the housings.

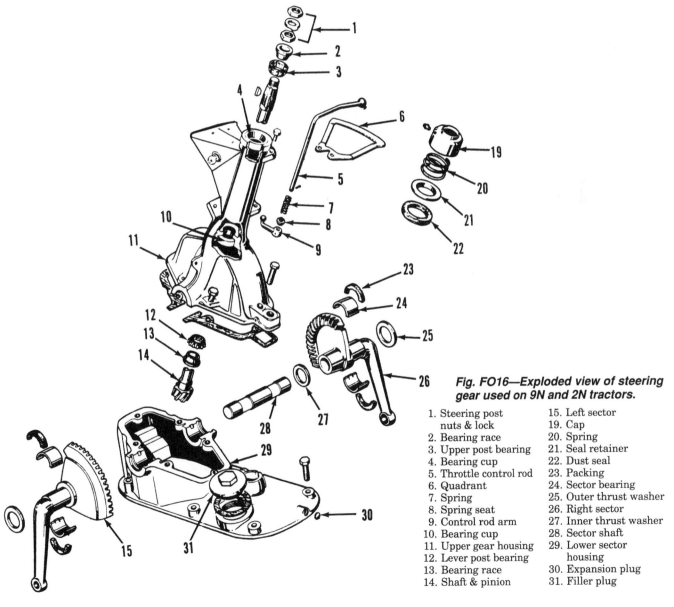

Fig. FO16—Exploded view of steering gear used on 9N and 2N tractors.

1. Steering post nuts & lock
2. Bearing race
3. Upper post bearing
4. Bearing cup
5. Throttle control rod
6. Quadrant
7. Spring
8. Spring seat
9. Control rod arm
10. Bearing cup
11. Upper gear housing
12. Lever post bearing
13. Bearing race
14. Shaft & pinion
15. Left sector
19. Cap
20. Spring
21. Seal retainer
22. Dust seal
23. Packing
24. Sector bearing
25. Outer thrust washer
26. Right sector
27. Inner thrust washer
28. Sector shaft
29. Lower sector housing
30. Expansion plug
31. Filler plug

19

To check the steering adjustment, first disconnect drag links from steering arms and turn steering wheel to the mid or straight ahead position. (Steering sector arms should be parallel and both point slightly rearward when in straight ahead position.) Remove the four cap screws on each side of steering housing which retain the sector shaft housings (Fig. FO18). Obtain maximum clearance for sector shafts and ball nut by rotating sector shaft housings, then check steering wheel shaft bearings by pulling up and pushing down on steering wheel. If looseness is present, adjust worm shaft bearings to a slight preload by

varying thickness of shims (3—Fig. FO17) between steering shaft tube and steering gear housing. (Removing shims decreases end play.) Be sure that worm shaft turns freely. If unable to adjust worm shaft end play, overhaul steering gear as outlined in paragraph 17.

If steering shaft bearings are satisfactory, check sector mesh as follows: Hold steering wheel steady to prevent movement of worm, grasp left sector arm and check gear backlash between sector and ball nut. If any backlash can be felt, adjust mesh position.

To adjust sector mesh, remove four cap screws from left sector shaft housing (if not already removed) and rotate sector housing forward (counterclockwise) until backlash is removed between left sector shaft and ball nut. Align four holes in left sector shaft cover, gasket and housing and reinstall retaining cap screws. Remove four cap screws from right sector shaft housing and rotate right sector shaft housing

Fig. FO17—Cross-sectional side and top views of steering gear unit used on 8N tractors prior to serial number 216989. Sectors (6 and 7) have straight teeth and left sector (6) is meshed with ball nut.

1. Shaft upper bearing
2. Shaft & worm
3. Shims
4. Worm bearings
5. Ball nut
6. Left sector
7. Right sector

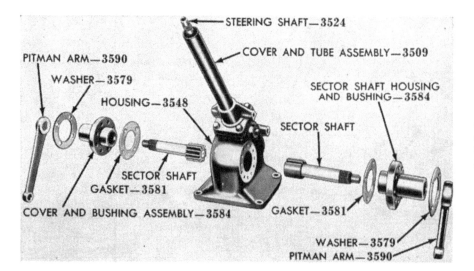

Fig. FO18—Disassembled sector shafts of steering gear used on 8N tractors prior to serial number 216989.

STEERING SHAFT—3524
COVER AND TUBE ASSEMBLY—3509
PITMAN ARM—3590
WASHER—3579
HOUSING—3548
SECTOR SHAFT HOUSING AND BUSHING—3584
SECTOR SHAFT
SECTOR SHAFT GASKET—3581
COVER AND BUSHING ASSEMBLY—3584
GASKET—3581
WASHER—3579
PITMAN ARM—3590

forward (clockwise) until backlash is removed and cap screw holes are aligned. Reinstall retaining cap screws. If excess clearance (backlash) still remains in steering, overhaul the unit as outlined in paragraph 17.

17. R&R AND OVERHAUL. To overhaul the steering gear on early 8N models, proceed as follows: Shut off the fuel and disconnect the fuel line at the carburetor, then remove the hood and fuel tank as a unit. Remove the battery and instrument panel. Disconnect drag links at steering gear arms. Unbolt steering gear housing from transmission housing and lift the unit off the tractor.

Pull both steering gear arms off sector shafts, then unbolt and remove sector shafts and housings from sides of gear case (Fig. FO18). (Turn housings clockwise as they are withdrawn.) Remove steering wheel shaft tube flange cap screws and lift the assembly off steering gear case.

> **CAUTION: Do not turn worm shaft if ball nut is near either end of worm, or ball retainers may be damaged.**

Separate the cover and tube assembly (3509—Fig. FO19) from steering worm shaft (3524). Inspect bearings (3517 and 3571) and seals (3518 and 3570) and renew if needed.

> **NOTE: Do not disassemble the ball nut (3539) and worm shaft (3524) shown in Fig. FO19. Any damage or excessive wear in any of the components of screw shaft and ball nut assembly is corrected by renewal of the unit, as individual parts are not available.**

New sector shaft housing bushings are pressed into housing 1/8 inch (3 mm) below face of hub. Outer bushings are installed flush or slightly below bottom of dust seal counterbores. Bushings should be align reamed to 1.125-1.126 inches (28.58-28.60 mm).

Install worm shaft and nut assembly with nut positioned at approximate center of worm, and adjust worm shaft bearings to a slight preload by means of

shims (3—Fig. FO17) between steering shaft tube and gear housing. Be sure that worm shaft turns freely. After bearing adjustment is completed, hold ball nut and rotate worm shaft until nut center tooth aligns with center of left sector shaft opening. Assemble left sector shaft (three large and four small teeth) and its housing to gear housing, with center tooth of three large sector teeth meshed with center space of teeth on ball nut as shown in Fig. FO17. The sector housing is installed with locating notch at bottom. Adjust left sector backlash as described in paragraph 16. Assemble right sector, meshing center tooth with third tooth space on left sector gear, counting from solid section of gear located on bottom as shown in Fig. FO17.

> **NOTE: Right sector center tooth is marked on end of tooth and left sector tooth space is similarly marked.**

Adjust backlash as described in paragraph 16. Fill gear housing with SAE 90 gear lubricant through filler hole in side of housing and reinstall assembly.

Models 8N After Serial Number 216988
The steering gear used on these tractors is a Saginaw screw and recirculating ball nut type, mounted on top of transmission housing. See Fig. FO20. The sector teeth are bevel type and right sector meshes with the ball nut. The left sector meshes with the right sector and rotates in the opposite direction.

18. ADJUSTMENT. To adjust the steering gear, first make certain that gear housing is properly filled with lubricant. Disconnect both drag links from steering gear arms to remove load from the gear unit and proceed as follows: To check wormshaft end play, first loosen the locknuts on the sector shaft adjusting screws (2—Fig. FO20) and back the screws out at least two full turns. Check steering wheel shaft bearings by pulling up and pushing down on steering wheel. If the end play of the worm shaft (steering wheel shaft) is not within the desired limits of 0.006-0.010 inch (0.15-0.25 mm), adjust the end play by

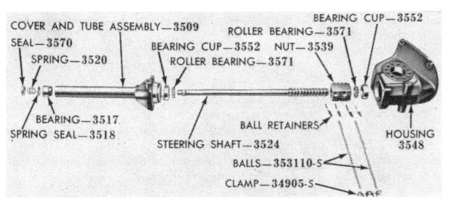

Fig. FO19—Disassembled steering column, worm shaft and ball nut used on 8N tractors prior to serial number 216989. Steering shaft (3524), nut (3539) and balls (353110-S) are not available separately.

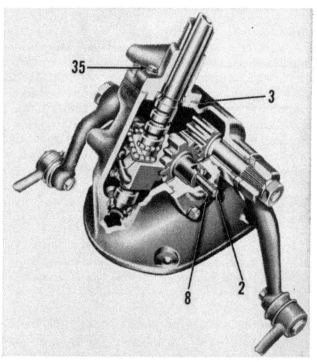

Fig. FO20—Cutaway view of steering gear used on 8N tractors after serial number 216988. Adjustment of worm shaft bearings is controlled by shims (3). Backlash of sector gear is controlled by a screw (2) on each side of the housing.

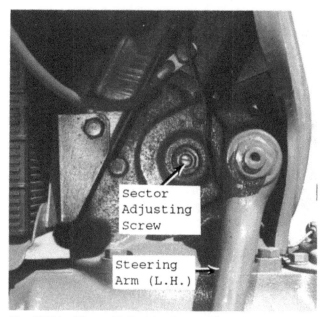

Fig. FO21—View of left side of steering gear.

Fig. FO22—View of right side of steering gear.

varying the thickness of the shim stack (3) between the steering shaft tube and the steering gear housing. Shims are available in various thicknesses. Ford recommends a minimum shim stack installation of not less than three 0.002 inch (0.05 mm) shims, or not less than two 0.005 inch (0.13 mm) shims. Tighten the steering shaft cover retaining cap screws to a torque of 25-30 ft.-lbs. (34-40 N·m). Renew worm shaft bearings if end play is over 0.010 inch (0.25 mm) with minimum recommended shim stack thickness.

After checking or adjusting worm shaft end play, adjust sector shaft end play as follows: Turn the steering wheel to the mid or straight ahead position. With the locknuts on both sector shaft adjusting screws (Fig. FO21 and FO22) loosened and the adjusting screw on the right side (as viewed from rear of tractor) backed out several turns, turn the adjusting screw on **left side** (Fig. FO21) of steering housing in (clockwise) until there is no perceptible end play in the sector shaft to which the **right steering arm** (Fig. FO22) is attached. While holding the adjusting screw in this position, tighten the locknut. Turn the adjusting screw on **right side** (Fig. FO22) of housing in until there is no perceptible end play in the sector shaft to which the **left steering arm** (Fig. FO21) is attached. Hold the adjusting screw in this position and tighten the locknut. Reconnect the drag links to the steering arms.

19. REMOVE AND REINSTALL. To remove the steering gear and housing assembly, first remove the steering wheel and withdraw the spring, felt packing and spring seat from the top of the steering column. Shut off the fuel and disconnect fuel line from the carburetor. Remove the hood and fuel tank. Disconnect throttle rod bracket from the transmission. Disconnect the tachometer cable, ammeter lead wire and oil pressure gauge line at instrument panel. Discon-

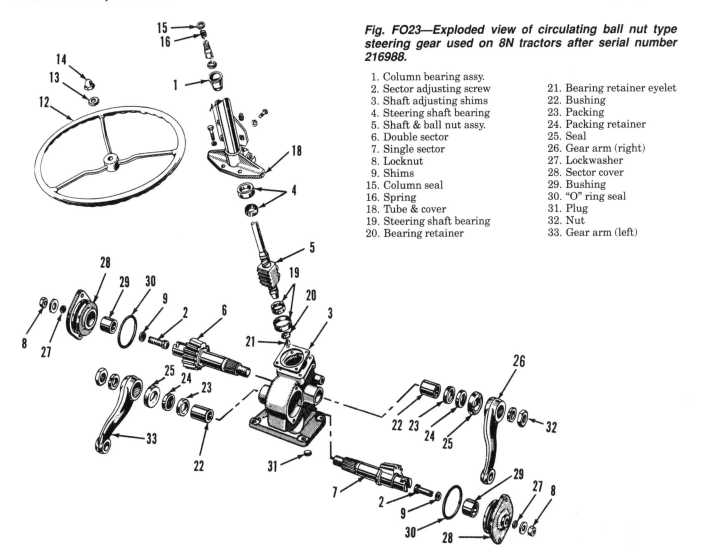

Fig. FO23—Exploded view of circulating ball nut type steering gear used on 8N tractors after serial number 216988.

1. Column bearing assy.
2. Sector adjusting screw
3. Shaft adjusting shims
4. Steering shaft bearing
5. Shaft & ball nut assy.
6. Double sector
7. Single sector
8. Locknut
9. Shims
15. Column seal
16. Spring
18. Tube & cover
19. Steering shaft bearing
20. Bearing retainer

21. Bearing retainer eyelet
22. Bushing
23. Packing
24. Packing retainer
25. Seal
26. Gear arm (right)
27. Lockwasher
28. Sector cover
29. Bushing
30. "O" ring seal
31. Plug
32. Nut
33. Gear arm (left)

nect the battery ground cable from steering gear housing and wires from junction block on steering column. Remove the generator regulator from bracket on steering column. Unbolt instrument panel and slide the panel over top of the steering column. Unbolt battery carrier from steering gear housing. Disconnect head light switch (if so equipped) and ignition switch from the hood rear lower panel Disconnect drag links from pitman arms. Remove the cap screws attaching steering gear housing to transmission case and lift the steering gear assembly from the tractor.

Installation is the reverse of removal. Be sure that steering gear is in the midposition and front wheels are positioned straight ahead when reconnecting drag links to the pitman arms.

20. OVERHAUL. To disassemble the steering gear, remove the pitman arm retaining nuts and pull pitman arms (26 and 33—Fig. FO23) from sector shafts (6 and 7). Unbolt the sector shaft side covers

(28) and remove the adjusting screw lock nuts (8). Turn the adjusting screws (2) in and remove the side covers (28) and sector shafts (6 and 7). Unbolt steering housing upper cover (18) from housing and remove cover, shims, shaft and ball nut assembly.

NOTE: Do not disassemble the ball nut assembly (5) as component replacement parts are not available. If the steering shaft and/or ball nut are worn or damaged, renew the complete assembly.

The renewable bushings (22—Fig. FO23) in housing have a bore diameter of 1.245-1.250 inches (31.63-31.75 mm). Bushings (29) in side covers have a bore diameter of 1.1255-1.1260 inches (28.59-28.60 mm). Renew packing (23, 24 and 25) in side covers. Shims (9) on the adjusting screws (2) are available in various thicknesses. When reassembling, use a shim of proper thickness to provide zero to 0.002 inch (0.05 mm) clearance between adjusting screw head and slot in sector shafts. Center the ball nut on worm shaft and insert shaft in housing. Bolt the housing upper (steering shaft) cover in position using the necessary number of shims (3) to provide 0.006-0.010 inch (0.15-0.0.25 mm) end play of worm shaft. Tighten the cover retaining cap screws to a torque of 25-30 ft.-lbs. (34-40 N•m).

NOTE: Ford recommends a minimum installation of two 0.005 inch (0.13 mm) shims (3—Fig. FO23), or three 0.002 inch (0.05 mm) shims for proper sealing of cover to steering gear housing. If worm shaft end play is over 0.010 inch (0.25 mm) with the minimum number of shims installed, renew the worm shaft bearings.

Assemble the sector shafts (6 and 7—Fig. FO23) and their adjusting screws (2) to the side covers (28), using the screws to pull the shafts into cover bushings. Center the ball nut rack on the steering worm, then install the double sector shaft (6) and cover into the housing with the one blocked tooth facing upward (Fig. FO20). The middle tooth on this sector shaft must mesh with the middle groove in the ball nut rack. Install the single sector shaft (7—Fig. FO23) and cover, meshing the fourth tooth of this shaft with the fourth groove in the left sector shaft previously installed. Tighten the side cover retaining cap screws to a torque of 25-30 ft.-lbs. (34-40 N•m). Refill the steering gear housing with SAE 90 gear lubricant.

Adjust the sector shaft end play as outlined in paragraph 18, and tighten the adjusting screw locknuts. Turn steering gear to mid or straight ahead position and install steering arms.

ENGINE AND COMPONENTS

R&R ENGINE WITH CLUTCH

21. All models are equipped with a four-cylinder, L-head engine. See Fig. FO24. A small number of tractors, identified as "NAN" models (for example: 9NAN), were built to operate on kerosene. Other than the fuel system components and a lower compression ratio of 4.75:1, the kerosene and gasoline engines are mechanically similar. Most kerosene burning models have since been converted to gasoline.

The engine and clutch assembly may be removed as a unit as follows: Drain cooling system and oil pan. Remove hood and fuel tank. Disconnect battery ground strap and cable at starter. Disconnect distributor and generator wires at front end and swing wiring out of the way. Disconnect throttle rod and choke control linkage. Disconnect the Proofmeter cable, ammeter lead wire and oil pressure gauge line. Disconnect the radiator hoses and the carburetor air intake hose. Block up under transmission and support engine with a chain hoist. See Fig. FO25. Disconnect front end of radius rod and drag link from either side of front axle. Remove cap screws attaching axle front support to the oil pan, then swing axle and

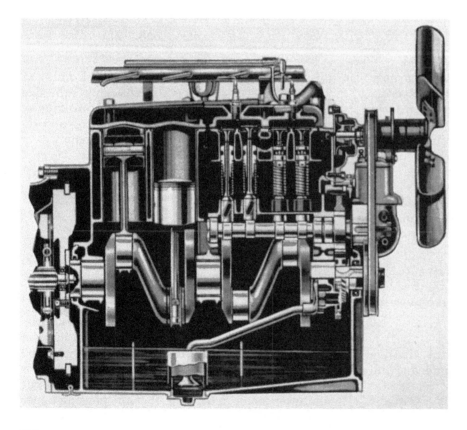

Fig. FO24—Cross section view of engine used in Models 9N, 2N and 8N.

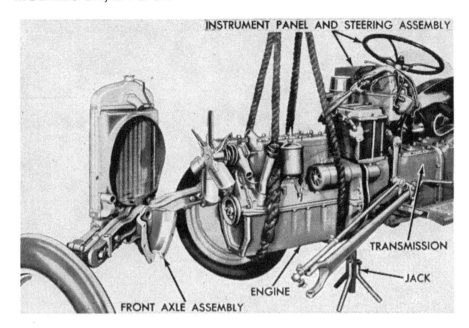

Fig. FO25—Method of removing engine on 9N, 2N and 8N tractors.

front end assembly away from the engine. Remove the cap screws attaching engine to transmission housing, then separate the engine from transmission.

Reinstall engine in reverse order of removal while noting the following special instructions. If the clutch was removed from the flywheel, use a dummy transmission clutch shaft or other suitable alignment tool to center the clutch disc on the flywheel when installing clutch on the flywheel. The engine should slide easily up against the transmission housing when properly aligned.

COMPRESSION PRESSURE

22. A compression test measures the compression pressure built up in each cylinder. The results can be used to assess general cylinder and valve condition. In addition, it can warn of developing problems inside the engine.

Start the engine and warm to normal operating temperature (upper radiator hose hot). Shut the engine off. Make sure the choke and throttle valves are wide open. Remove all spark plugs. Connect a remote starter switch to the starter relay according to manufacturer's instructions. Leave the ignition switch in the OFF position. Connect a compression gauge to the No. 1 cylinder following the manufacturer's instructions. Crank the engine at least five compression strokes with the remote starter switch or until there is no further increase in compression shown on the tester gauge. Remove the compression tester and record the reading. Repeat the operation to test each remaining cylinder.

Minimum compression pressure is 90 psi (620 kPa). The lowest pressure reading must be within 75 percent of the highest. A greater difference indicates

worn or broken rings, leaking or sticking valves or a combination of these problems.

If the compression test indicates a problem, isolate the cause with a wet compression test. This is done in the same way as the dry compression test already described except that about one tablespoon of engine oil is poured down the spark plug hole before performing the second test.

If the wet compression readings are much greater than the dry readings, the trouble is probably caused by worn or broken piston rings. If there is little difference between the two readings, the problem is probably due to leaking or sticking valves. When two adjacent cylinders have similar low readings during both dry and wet testing, the problem may be a defective head gasket between the two cylinders.

CYLINDER HEAD

23. REMOVE AND REINSTALL. Remove hood and fuel tank as a unit. Drain the cooling system and disconnect the upper radiator hose. Remove the ignition cable harness and the spark plugs. Remove the cap screws attaching the oil filter housing and the ignition coil (models with angle drive distributor) to the cylinder head. Remove the cylinder head stud nuts or bolts, then remove the cylinder head and gasket.

Thoroughly clean the cylinder head and cylinder block surfaces. Check the cylinder head for distortion, cracks or other damage and resurface or renew the cylinder head if necessary.

Reinstall the cylinder head using a new gasket. Gradually tighten the cylinder head stud nuts or bolts in three stages. Begin tightening nuts or bolts located in the center of the cylinder head and tighten alternately toward each end. A recommended tight-

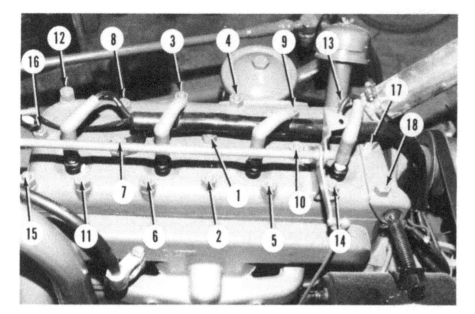

Fig. FO26—Recommended cylinder head bolt tightening sequence.

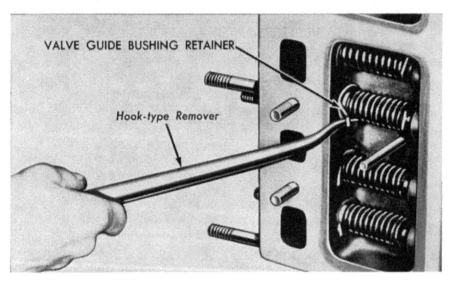

Fig. FO27—Removing valve guide retainer preparatory to removal of valve, spring and guide as an assembly.

ening sequence is shown in Fig. FO26. Final tightening torque is 50-55 ft.-lbs. (68-74 N·m) for stud nuts and 65-70 ft.-lbs. (88-95 N·m) for bolts.

VALVES

Intake and Plain Exhaust

24. Intake and exhaust valves are not interchangeable. Valves removed for grinding or refacing must be reinstalled in their original location. Valves, guides and springs can be removed from the top of the cylinder block after first removing the cylinder head, intake and exhaust manifold, valve chamber covers and valve guide retainers. Use a hook-type puller as shown in Fig. FO27 to extract retainers from valve guides, then remove each valve, guide and spring from cylinder block with a jack type lifter as shown in Fig. FO28. Keep valves and other parts identified in their removal order for proper reassembly.

Renew valves having deeply pitted or warped heads, or stems that are bent, scored or worn. Reface and reseat intake and exhaust valves to an angle of 45°. Valve specifications are as follows:

Valve Face Angle	45°
Valve Seat Angle	45°

Valve Seat Width
Intake . . . . 0.060-0.080 in.
(1.5-2.0 mm)
Exhaust. . . . . 0.070-0.090 in.
(1.8-2.3 mm)

Valve Stem Diameter
Two-piece Guide. . . . . 0.3105-0.3115 in.
(7.887-7.912 mm)
One-piece Guide. . . . . 0.341-0.342 in.
(8.662-8.686 mm)

Clearance In Guide
Intake . . . . 0.002-0.004 in.
(0.051-0.101 mm)

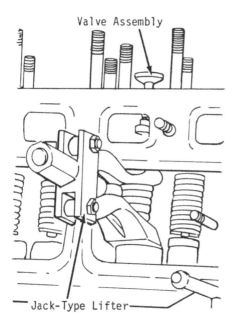

Fig. FO28—Removing valve spring and guide as an assembly.

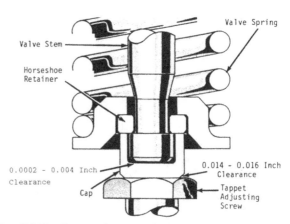

Fig. FO30—Free valve type exhaust valve rotators are used on late production engines and are available in kit form for installation on early engines.

Exhaust .... 0.0025-0.004 in.
(0.064-0.101 mm)

Use a lever type tool to reinstall valve guide retainers as shown in Fig. FO29. Correct valve tappet clearance (engine cold) is 0.010-0.012 inch (0.26-0.30 mm) for intake and 0.014-0.016 inch (0.36-0.40 mm) for exhaust valves. Refer to paragraph 28 for valve tappet adjustment procedure. On engines without adjustable type tappets, grind end of valve stem to correct insufficient clearance, or seat faces to correct excessive clearance.

Free Type Exhaust Valves and Rotators

25. Some 8N engines were originally equipped with free valve type exhaust valve rotators which are also available as a kit for engines not so equipped. The kit contains among other items: eight adjustable type tappets (push rods), four exhaust valves equipped with special caps and horseshoe type spring retainers as shown in Fig. FO30.

The procedure for removing and refacing the free type exhaust valves is identical to the procedure outlined for the plain exhaust valves in paragraph 24. Tappet clearance (engine cold) of 0.010-0.012 inch (0.26-0.30 mm) for intake valves and 0.014-0.016 inch (0.36-0.40 mm) for exhaust valves is the same as on engines without rotator caps.

The free valve type rotators and valves will not function properly unless there is clearance (end play) between the end of the exhaust valve stem and the inside floor of the cap when the open end of the cap just contacts the spring keeper horseshoe as shown in Fig. FO30. Desired end gap is 0.0002-0.004 inch (0.005-0.10 mm).

Rotator cap end clearance should be checked and adjusted, if necessary, each time the valves are reseated. One method of checking this clearance using a dial indicator is shown in Fig. FO31. Turn crankshaft until exhaust valve is in fully open position. Position dial indicator on valve head, then move valve vertically with your fingers and observe indicator reading. If clearance is more than 0.004 inch (0.10 mm), remove the rotator cap and reduce its length by

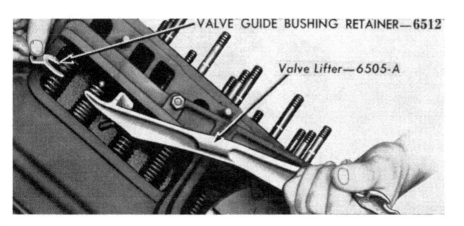

Fig. FO29—Special tool for installing valve guide retainers.

Fig. FO31—With valve on lift portion of cam, the rotator cap gap can be measured using a dial indicator as shown. If gap is correct, valve will have up and down free play or travel of 0.0002-0.004 inch (0.005-0.10 mm).

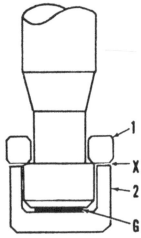

Fig. FO32—Feeler gauge and shim stock method of checking end clearance between rotator cap and end of valve stem. Refer to text.

1. Horseshoe keeper
2. Rotator cap
G. 0.010 in. (0.25 mm) shim stock
X. Gap between rotator cap & keeper

lapping open end of cap on emery cloth laid on a flat, smooth surface. If clearance is less than 0.0002 inch (0.005 mm), install new horseshoe retainer; or if retainer is not worn, grind end of valve stem until specified clearance is obtained.

An alternate method of checking the end gap is as follows: Cut a ³⁄₁₆ inch (4.5 mm) diameter disc from a strip of 0.010 inch (0.25 mm) thick shim stock. Place this disc (G—Fig. FO32), which must be flat, inside the rotator cap (2). Install the horseshoe keeper (1) on the valve stem (without the spring). Now, while simultaneously pressing downward on the horseshoe keeper and upward on the rotator cap (2), measure the gap (X) between the cap and keeper with a feeler gauge. If the gap measures anywhere between 0.006 inch (0.15 mm) and 0.009 inch (0.23 mm), it is within the desired limits. If the measured gap is less than 0.006 inch (0.15 mm), reduce the length of the cap by

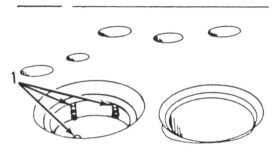

Fig. FO33—An alternate method of removing valve seat inserts is to electric weld three small beads (1) on insert as shown. Lift insert out after it has cooled.

grinding or lapping the open end face of the cap. If the measured gap is more than 0.009 inch (0.23 mm), renew the valve cap and/or horseshoe keeper, or grind the end of the valve stem.

VALVE SEATS

26. All tractors except 2N within the serial number range of 86271 to 168259 were originally equipped with intake and exhaust valve seats of the hardened insert type. Model 2N tractors within the preceding serial number range were equipped only with exhaust valve seat inserts.

If valve seat renewal is necessary and special equipment for removal of the seat inserts is not available, the following alternate method of removal may be used. Use an electric welder to weld three small beads on the insert as shown in Fig. FO33. Be careful not to allow the welding rod to contact the cylinder block. The seat insert will shrink as it cools and can then be easily lifted or pried from the cylinder block counterbore.

The counterbore in the cylinder block must be 0.0015-0.003 inch (0.038-0.076 mm) smaller than the measured diameter of the new seat insert. Cool the new insert in dry ice or a freezer for at least 15 minutes, then drive or press the insert into the counterbore using a suitable installing tool. Make certain that the insert is bottomed squarely in the counterbore. After the insert is installed, grind the seat to a 45° angle using a seating stone or cutter. If a seat cutter is used, each valve should be lapped in its seat. Seat width should not exceed 0.125 inch (3.2 mm). If the seat requires narrowing, use a 30° stone for the top of the seat and a 60° stone for the bottom.

VALVE GUIDES AND SPRINGS

27. Early production valve guides used in 9N, 2N and 8N (prior to serial number 42162) tractors are made in halves. The guide halves should be kept together with their mating valves to maintain stem

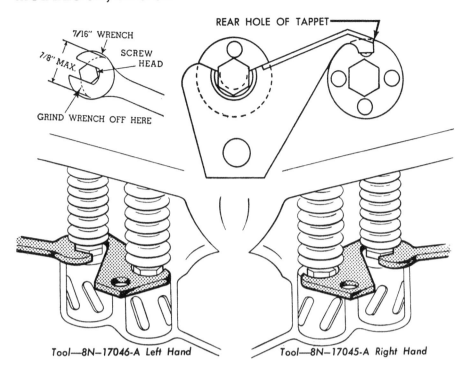

Fig. FO34—Special tappet holding wrenches are used for adjustment of valve tappets. A regular end wrench should be ground off as shown in upper left.

to guide clearance if guides are reused. Later production guides used in 8N tractors after serial number 42161 are one-piece type.

Valves used with two-piece guides have a stem diameter of 0.311 inch (7.90 mm). Stem diameter of valves used with one-piece guides is 0.341 inch (8.66 mm). Stem to guide clearance wear limit is 0.005 inch (0.13 mm) for intake and 0.006 inch (0.15 mm) for exhaust valves for either type guide.

The one-piece valve guides can be installed in engines originally equipped with the two-piece type. If valve rotators are not included in the change-over, it will be necessary to use the same valve springs as used for the two-piece type of guide.

Valve springs in tractors with two-piece valve guides should test 37-40 pounds (165-180 N) at 2.125 inches (54 mm); springs for one-piece guides should test 41-44 pounds (182-195 N) at 1.80 inches (45.7 mm). Renew valve springs if protective paint coating is lost or if tension is less than specified.

VALVE PUSH RODS (TAPPETS)

28. Valve push rods (tappets) are of the barrel type. On tractors built prior to 1951, the tappets were of the nonadjustable type; later tractors were equipped with adjustable tappets. Adjustable type tappets can be installed in engines originally equipped with the nonadjustable type, and this change-over is recommended. Adjustable tappets were also included in Ford Free Type Valve and Tappet Kit (8N-6546B) which provided special exhaust valves and free type exhaust valve rotators. The kit could be installed in any 9N, 2N or 8N tractor.

Tappet clearances (gap) should be 0.010-0.012 inch (0.26-0.30 mm) for intake and 0.014-0.016 inch (0.36-0.40 mm) for exhaust with the engine cold. The method of adjusting the tappets is shown in Fig. FO34. Required are a left-hand tappet holding wrench (Ford 8N-17046A), right hand tappet holding wrench (Ford 8N-17045A) and a conventional tappet wrench ground off so that it does not extend past the adjusting screw.

To make the adjustment, remove the tappet chamber covers from right side of engine. Rotate the crankshaft until the number one (front) piston is at top center on compression stroke (both valves closed). Insert protruding tip of tappet holding wrench into rear hole of the tappet being adjusted and the other end of the wrench over the head of the adjacent tappet. Measure the clearance between tappet adjusting screw head and valve stem cap using a feeler gauge. Turn the tappet adjusting screw as necessary to obtain the recommended clearance. Repeat the adjustment procedure for valves of each cylinder, following the firing order of 1-2-4-3.

NOTE: Turning the crankshaft 180° (½ revolution) will position the next piston in the firing order at top center of compression stroke.

TIMING GEARS AND COVER

29. REMOVE AND REINSTALL. Shut off the fuel, disconnect fuel line at the tank and remove the hood and fuel tank. Drain the cooling system and disconnect the upper radiator hose. Block up under front end of engine behind axle support and remove the bolts holding the axle support in place. Discon-

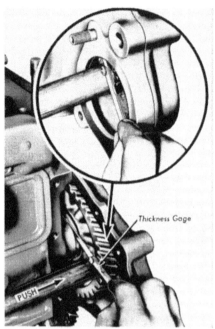

Fig. FO35—If camshaft end play exceeds 0.004 inch (0.10 mm) when checked as shown, install a thinner timing gear cover gasket or renew the cover.

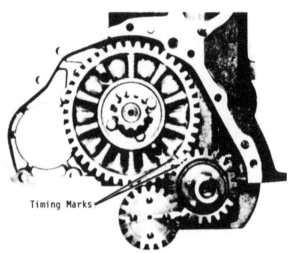

Fig. FO36—View of crankshaft gear and camshaft gear with valve timing marks properly aligned.

nect the front end of drag link and mating radius rod on one side of the tractor and swing the front end assembly away from engine as shown in Fig. FO25. Remove the distributor, fan and generator on tractors prior to serial number 263844; on later tractors, generator removal is not necessary. Remove hand crank jaw and crankshaft fan belt pulley. Remove the gear cover cap screws and pull cover off engine.

NOTE: On 8N tractors after serial number 263844, the timing gear cover controls the camshaft end play. With the governor assembly removed and gear cover in place, end play can be checked by measuring clearance between rear of camshaft gear flange and boss on engine block as shown in Fig. FO35.

Installation of the timing gear cover is the reverse of removal steps. If camshaft end play exceeds 0.004 inch (0.10 mm) on 8N tractors with angle mount distributor, a thinner cover mounting gasket should be used or cover should be renewed.

Camshaft Gear

30. The camshaft timing gear is made of aluminum alloy material. Backlash between camshaft gear and crankshaft gear should be 0.002-0.006 inch (0.05-0.15 mm).

On early Model 9N tractors, the gear is tightly pressed on the camshaft and requires special pullers and fixtures for proper removal and installation. On these tractors, the gear is usually removed with the camshaft as an assembled unit, then pressed or pulled off the camshaft.

Camshaft gears of the pressed-on type are supplied for service; however, service camshafts for this type of gear installation are discontinued, and only camshafts of the bolted-on gear type are supplied.

On later tractors, the camshaft gear is bolted onto the camshaft and may be renewed without removing the camshaft. The bolted-on type gear is available in 0.006 inch (0.15 mm) oversize to facilitate obtaining the recommended gear backlash.

When installing camshaft gear, mesh timing marks on gears as shown in Fig. FO36 and use care so as not to damage the comparatively soft tooth faces of camshaft gear.

Crankshaft Gear

31. Renewal of crankshaft gear requires removal of timing gear cover, oil pan and the front main bearing cap and oil pump assembly. Use a suitable puller to remove the gear while protecting end of crankshaft from damage. Heating new crankshaft gear in hot oil or water will facilitate installation. Be sure that timing marks on crankshaft and camshaft gear are aligned as shown in Fig. FO36.

CAMSHAFT

32. To remove the camshaft, first remove timing gear cover, cylinder head and valves as previously outlined. Lift the tappets out of the cylinder block bores. Pull camshaft forward from the cylinder block.

The camshaft is supported in unbushed bores in the cylinder block. Renew the camshaft if bearing journals or cam lobes are scored, pitted or excessively worn. For service, only the bolted-on gear type of camshaft is available. If the camshaft is renewed, the tappets should also be renewed.

Refer to the following specifications:

Camshaft Journal Diameter
Recommended . . . . 1.7965-1.7970 in.
(45.631-45.644 mm)
Wear Limit. . . . . 1.795 in.
(45.59 mm)

Bearing Clearance
Recommended . . . . 0.001-0.002 in.
(0.025-0.051 mm)
Wear Limit. . . . . 0.004 in.
(0.10 mm)

End play
Recommended . . . . 0.0015-0.004 in.
(0.038-0.10 mm)

Lubricate the camshaft lobes and journals prior to installation. Install the camshaft, making certain that timing gear marks are aligned as shown in Fig. FO36. On 8N tractors with angle mount distributor, check the camshaft end play by measuring clearance between rear of camshaft gear flange and boss on engine block through governor opening, or after removing side gear cover as shown in Fig. FO35. If more than 0.004 inch (0.10 mm) clearance exists, use a thinner timing gear cover gasket or renew the cover. Check valve tappet gap and adjust if necessary as outlined in paragraph 28.

CONNECTING ROD AND PISTON UNITS

33. Piston and connecting rod assemblies may be removed from the top after cylinder head, oil pan and connecting rod bearing caps are removed. Prior to removal, identify all unmarked piston and connecting rod units to correspond with the cylinder in which they are installed. All components should be reinstalled in their original positions if reused.

NOTE: Remove ring ridge and carbon deposits from top of cylinder before pushing the pistons out of the cylinder block.

Install pistons and connecting rods so that notch in piston crown is toward the front of the engine. The oil squirt hole in the connecting rod can face to the front or the rear. Lubricate pistons and cylinder walls with engine oil and use a suitable ring compressor to compress the piston rings. Tap the piston down into the cylinder, being careful not to allow the connecting rod studs to scratch the crankshaft journal.

NOTE: Pieces of plastic or rubber tubing may be placed over the connecting rod studs during piston installation to prevent damage to the crankshaft.

Apply a light coat of engine oil to connecting rod bearing inserts. Install bearing cap on the connecting rods, making sure that the number on the bearing cap

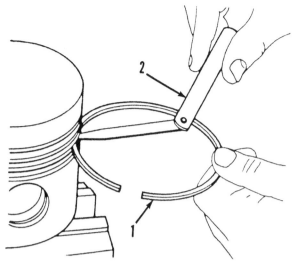

Fig. FO37—Insert a new piston ring in piston ring groove and measure side clearance as shown to determine if ring grooves are worn.

is on the same side as the number on the connecting rod. Tighten connecting rod nuts evenly to a torque of 35-40 ft.-lbs, then install new locknuts (pal nuts) and tighten to 4-5 ft.-lbs. torque.

CAUTION: Connecting rod locknuts should not be reinstalled after they have been used once.

PISTON RINGS

34. Two compression rings and one oil control ring are used on each piston. New rings are marked to indicate top side and must be installed accordingly. Counterbore on the top ring must face upward. The second ring is provided with an expander.

Thoroughly clean the piston ring grooves, then check grooves for wear as follows. Position new rings in piston grooves and measure side clearance between each ring and piston lands using a feeler gauge as shown in Fig. FO37. Excessive side clearance will allow the rings to "flutter" in the piston grooves when the engine is running, resulting in poor sealing of the combustion chamber and eventual ring breakage. Renew piston if ring side clearance exceeds 0.004 inch (0.10 mm).

Before installing new rings on a piston, check the rings for proper end gap as follows. Each ring should be fitted and checked in the cylinder in which it is going to be used. Push the ring into the cylinder using an inverted piston to ensure that the ring is square with the cylinder wall. Measure the gap between the ends of the ring with a feeler gauge as shown in Fig. FO38.

CAUTION: It is important that all rings have at least the minimum end clearance to allow for expan-

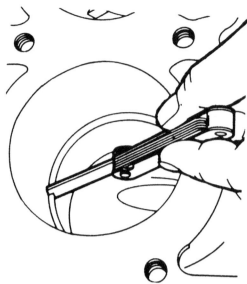

Fig. FO38—Push new piston rings squarely into the cylinder and measure ring end gap as shown.

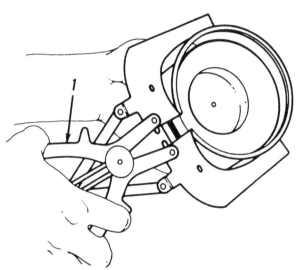

Fig. FO39—Use a suitable expander tool when installing piston rings to prevent overexpansion and distortion of the rings.

sion that occurs when engine reaches operating temperature. Otherwise, the ring ends may butt together and cause scuffing, scoring and ring breakage.

The ends of the ring may be filed, if necessary, to obtain the recommended end gap. It is recommended that a suitable piston ring expander tool be used when installing the piston rings on the piston (Fig. FO39). Use of an expander tool will prevent over-expansion and possible distortion of the rings. Stagger the piston ring end gaps around the piston prior to installation of piston in the cylinder block.

Refer to the following specifications:

Piston Ring End Gap . . . . 0.010-0.017 in.
(0.25-0.43 mm)

Piston Ring Side Clearance
Top Ring . . . . 0.0015-0.0035 in.
(0.038-0.089 mm)
Second Ring . . . . 0.0010-0.0025 in.
(0.025-0.063 mm)
Oil Control Ring . . . . 0.0015-0.0030 in.
(0.038-0.076 mm)

PISTON PINS

35. The full floating piston pin is retained in the piston by spring-steel lock rings, which should be renewed if they are removed for any reason.

Piston pin diameter is 0.750 inch (19.05 mm). New piston pins are available only in a set with a new piston and pin retainers. Pins for aluminum pistons are 2.844 inches (72.24 mm) long and are painted pink for identification. Pins for cast steel pistons are 2.972 inches (75.49 mm) long and are painted green for identification.

With both piston and pin clean and at the same temperature, the pin should be a thumb push fit in piston pin bore. Pin fit in rod bushing is correct if pin drops slowly through the bushing of its own weight.

SLEEVES AND PISTONS

36. Hardened steel sleeves were used in production up to tractor serial number 8N433578; cast iron sleeves were used in later production tractors. Engines with cast iron sleeves can be identified externally by the diamond stamped before and after the serial number, which replaces the star used on engines with steel sleeves.

The outside diameter of cast iron sleeves is approximately 0.098 inch (2.49 mm) larger than steel sleeves, hence, the diameter of the bores in the cylinder block equipped with cast iron sleeves is about 0.098 inch (2.49 mm) larger than the block bore of engines equipped with steel sleeves. Cast iron sleeves are also flanged at the top. Installation of cast iron sleeves in a block originally equipped with steel sleeves necessitates the reboring and counterboring of the cylinder block.

Bare pistons are not available, but pistons with fitted pins and pin retainers are furnished in standard size and 0.020 inch (0.50 mm) oversize.

Check pistons and sleeves for scoring or other damage. Check sleeves for out-of-round and taper (wear) by measuring at four different locations as shown in Fig. FO40. Refer to the following specifications:

Sleeve Inside Diameter . . . . 3.1875-3.1885 in.
(80.96-80.99 mm)

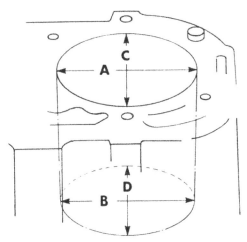

Fig. FO40—Use a bore gauge to check inside diameter of cylinder sleeves for taper or out-of-round.

Sleeve Out-Of-Round
 Wear limit . . . . 0.003 in.
                                              (0.076 mm)
Sleeve Taper
 Wear limit . . . . 0.006 in.
                                              (0.15 mm)

If the cylinder walls have only minor surface imperfections and the out-of-round and taper are within limits, it may be possible to remove the imperfections by honing the cylinder walls and installing new piston rings, providing the piston to cylinder clearance is within limits. Cylinder walls that are severely marked or worn beyond the specified limits should be bored for installation of oversize pistons, or new sleeves should be installed.

Steel sleeves are usually removed from the cylinder block with a crushing tool. Drive the tool to the bottom of the cylinder and lift out the collapsed

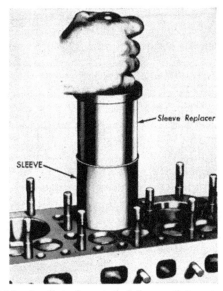

Fig. FO41—A piloted mandrel is used to prevent wall buckling of steel cylinder sleeves during installation.

sleeve. Install a new sleeve using a suitable piloted arbor as shown in Fig. FO41, then check installation with a bore gauge or a new piston and feeler ribbon of correct thickness. If recheck shows new sleeve was buckled during installation, it must be removed and a new one installed.

Cast iron sleeves should be removed from block using a sleeve puller. They should be carefully installed and checked for distortion after installation. If check shows sleeves have distorted or if piston skirt clearance is less than specified, sleeve should be final sized with a suitable cylinder hone.

Cast iron sleeves can be used for servicing an engine originally equipped with steel sleeves. When making this conversion, observe the following: First, measure the outside diameter of the cast iron sleeve, diameter of sleeve shoulder, and height of sleeve shoulder. Then adjust the boring equipment to provide the cast iron sleeve with a 0.001 inch (0.025 mm) press (interference) fit in block bore. Adjust the counterboring equipment to provide the sleeve shoulder with a 0.001-0.003 inch (0.025-0.076 mm) standout above top surface of cylinder block, and a 0.003-0.006 inch (0.076-0.15 mm) side clearance of sleeve shoulder in block counterbore. It is suggested that 0.060 inch (1.5 mm) on the diameter be removed for the first rough cut, and approximately 0.035 inch (0.89 mm) for the second rough cut. The finish cut should be made to provide the sleeve with a 0.001 inch (0.025 mm) press fit.

After installing the cast iron sleeve, use a suitable hone to final size the sleeve to correct any distortion and also to provide the correct piston skirt clearance.

37. FINAL SIZING OF SLEEVES AND FITTING PISTONS. The pistons must be individually fit to the cylinders. To check the clearance of a piston in a cylinder bore, use a feeler gauge ribbon ½ inch (13 mm) wide and of specified thickness (see the thickness table in Fig. FO42) that is long enough to cover the entire length of the piston. Attach the feeler gauge to a piston pull scale (Fig. FO43). Invert the piston and insert piston and feeler gauge into cylinder until piston is flush with top of cylinder block. Make sure that piston pin bore is parallel to the crankshaft and feeler ribbon is 90° from pin bore as shown in Fig. FO43. Slowly withdraw the feeler ribbon and note pull scale reading. Fit is correct when 5 to 10 pounds (22-44 N) pull is required to withdraw the feeler ribbon.

To final size the cylinders use a rigid type hone and No. 220 grit stones. A drill with a speed of 250 to 450 rpm should be used to drive the hone. The stones must be used dry to obtain the desired cylinder bore finish. Cover the crankshaft (if not removed) with shop towels while honing the cylinders.

CYLINDER BORE AND PISTON COMBINATION	FEELER GAUGE THICKNESS	POUNDS PULL
New Sleeve—New Piston	0.002 inch	5-10
New Sleeve—Used Piston	0.003 inch	5-10
Used Sleeve—New Piston	0.003 inch	5-10
Used Sleeve—Used Piston	0.004 inch	5-10

Fig. FO42—Feeler gauge dimensions for fitting pistons in cylinder bores.

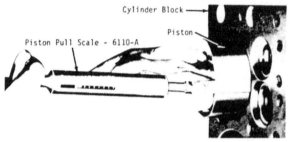

Fig. FO43—Piston fit is correct when 5 to 10 pounds (22-44 N) pull is required to withdraw feeler gauge of specified thickness as shown.

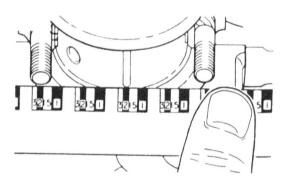

Fig. FO44—Connecting rod and main bearing clearance can be easily checked using Plastigage.

NOTE: The speed of the hone and rapidity of the stroke govern the crosshatch pattern in the bore. The crosshatch marks should intersect at approximately 90° for proper ring seating.

Operate the hone through the bore ten or twelve complete strokes. Remove the hone, clean the bore with dry shop towels, and recheck the piston fit as outlined above. Repeat the procedure until the desired piston to bore fit has been achieved.

38. CLEANING AFTER HONING. After the honing is completed, clean the cylinder block as follows: Wipe or remove as much of the abrasive material as possible from the cylinder walls using dry shop towels. Swab the cylinder wall with clean SAE 10 engine oil.

CAUTION: Do not use gasoline or kerosene to clean the sleeve walls after the honing operation. Solvents of this type will not remove the abrasive, but will further embed small particles in the pores of the sleeves.

Wipe the oil out of the cylinder with clean shop towels and repeat the procedure until the shop towel used to wipe out the oil appears clean. Wash the cylinder with hot, soapy water. Flush the water jackets to remove foreign material which might cause wear to the water pump. Remove the shop towels from the crankshaft and wash the crankshaft with hot soapy water. Dry the cylinder block thoroughly using compressed air. Apply clean engine oil to the cylinder walls and crankshaft to prevent rust.

CONNECTING RODS AND BEARINGS

39. Connecting rod bearings are slip-in precision type, held in position by lock tabs which fit into cutouts in connecting rod and cap bore. Bearing wear is corrected by installing new bearing shells which are made to close tolerance and do not require final sizing. New inserts are available in standard size and various undersizes.

If the bearing shells are scored, badly scratched, have the bearing overlay wiped out, show signs of fatigue failure or embedded dirt, install new bearing shells.

If the bearing shells appear to be serviceable, keep them with their respective connecting rods for reassembly in their original position in the engine.

Check the connecting rod bearing clearance with the crankshaft using Plastigage as follows: Place a piece of Plastigage of correct size across the full width of the bearing shell in the rod cap. Install the rod cap and tighten the retaining nuts to 35-40 ft.-lbs. (48-54 N.m) torque, being careful not to rotate the crankshaft while the Plastigage is in place. Remove the rod cap and measure the width of the flattened Plastigage using the Plastigage scale as shown in Fig. FO44. The width of the widest portion of the flattened Plastigage is the bearing clearance. If the bearing clearance is excessive, renew the bearing shells and/or regrind the crankshaft to permit the installation of appropriate undersize bearings.

Refer to the following specifications:

Crankpin Diameter . . . . 2.094 in.
(53.19 mm)

Bearing Clearance
  Desired . . . . 0.001-0.0025 in.
(0.025-0.063 mm)
  Wear Limit . . . . 0.005 in.
(0.13 mm)

Connecting Rod Side Clearance . . . . 0.004-0.011 in.
(0.10-0.28 mm)
Bearing Cap Nuts Torque. . . . . 35-40 ft.-lbs.
(48-54 N•m)

The piston pin bushing is renewable, use a suitable bushing driver and press to remove and install the pin bushings. The bushing oil hole, located by the oil hole in the connecting rod, is drilled after the new bushing is installed.

The connecting rods may be installed with the oil squirt hole facing forward or rearward. Coat connecting rod bearings with engine oil, then install bearing cap on the connecting rod. Make certain that the number on the bearing cap is on the same side as the number on the connecting rod. Tighten the connecting rod nuts evenly to a torque of 35-40 ft.-lbs. (48-54 N•m). Install new locknuts and tighten to 4-5 ft.-lbs. (5.5-6.5 N•m) torque.

> **CAUTION: Connecting rod locknuts should not be reused after once being installed and removed.**

Push the connecting rod to one side and measure the side clearance between the connecting rod and the crankshaft using a feeler gauge. Recommended side clearance is 0.004-0.011 inch (0.10-0.28 mm). Renew connecting rod if side clearance is excessive.

CRANKSHAFT AND MAIN BEARINGS

The crankshaft is supported by three main bearings of the nonadjustable, slip in precision shell type. Main bearing wear is corrected by installing new bearing shells. Bearing inserts are available in standard size and various undersizes.

40. REMOVAL. To remove the crankshaft, the engine must be removed from the tractor as previously outlined in paragraph 21. Remove the timing gear cover, oil pan, clutch and flywheel. Unbolt and remove connecting rod bearing caps and main bearing caps, making certain that all bearing caps are identified so they can be reinstalled in their original positions. Lift the crankshaft out of the cylinder block.

If bearing liners are scored, badly scratched, worn or show signs of fatigue failure or embedded dirt, install new bearing shells. If bearing liners appear to be serviceable, keep them with their respective bearing cap for reassembly in their original location in the engine.

41. INSPECTION. Thoroughly clean the crankshaft, including the drilled oil passages. Measure the diameter of each journal in four places to determine out-of-round, taper and wear (Fig. FO45). The amount of journal taper is the difference in measurements taken at points "A" and "B." The amount of

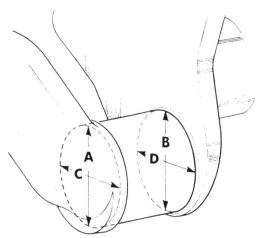

Fig. FO45—Measure crankshaft journal diameter in four locations to determine out-of-round, taper and wear.

journal out-of-round is the difference between measurements taken at "A" and "C," and at "B" and "D." If crankshaft main journals or crankpins are out-of-round more than 0.0015 inch (0.038 mm) or tapered more than 0.001 inch (0.025 mm), the shaft should be reground to the next undersize or renewed. If journals are worn evenly and not out-of-round or tapered more than the specified limit, undersize bearings may be installed without regrinding the journals, providing the bearing clearance will then be within specified limits.

Refer to the following specifications:
Main Journal Diameter
    Standard. 2.248-2.249 in.
(57.10-57.12 mm)
Main Bearing Running Clearance
    Desired . . . . 0.0010-0.0025 in.
(0.025-0.063 mm)
    Wear Limit . . . . 0.005 in.
(0.13 mm)
Crankpin Diameter
    Standard. . . . . 2.0935-2.0945 in.
(53.18-53.20 mm)
Rod Bearing Running Clearance
    Desired . . . . 0.001-0.0025 in.
(0.025-0.063 mm)
    Wear Limit . . . . 0.005 in.
(0.13 mm)
Main Bearing Nuts or Bolts Torque . . 90-100 ft.-lbs.
(122-135 N•m)

The main bearing clearance may be easily checked using Plastigage as follows: Place a piece of Plastigage of correct size across the full width of the bearing shell in the bearing cap. Install the bearing cap and tighten the retaining nuts or cap screws to a torque of 90-100 ft.-lbs. (122-135 N•m). Remove the bearing cap and measure the width of the flattened Plastigage using the Plastigage scale as shown in Fig. FO44. The width of the widest portion of the flattened Plastigage is the bearing clearance.

42. INSTALLATION. After bearing clearances are checked, lubricate the crankshaft journals and bearings with engine oil and install main bearing caps and bearings. Pry the crankshaft forward and push the center bearing cap rearward to align the thrust surfaces of the center main bearing.

> NOTE: In early production engines, the main bearing caps were held in place by studs and nuts. Later production engines use cap screws, which can be substituted for studs on engines originally so equipped, if desired.

Tighten main bearing cap nuts or cap screws evenly to a torque of 90-100 ft.-lbs. (122-135 N•m) Rotate the crankshaft during tightening to make sure it is not binding. If the crankshaft becomes hard to turn, stop and find out why before continuing. Check for foreign material on the bearings and journals. Make absolutely certain that the bearings are the correct size, especially if the crankshaft has been reground. Check the crankshaft end play with a dial indicator. Recommended end play is 0.002-0.006 inch (0.05-0.15 mm). If end play exceeds 0.008 inch (0.20 mm), renew the thrust bearing. If end play is less than specified limit, check for burrs or dirt on thrust surfaces or for misalignment of side flanges on center main bearing. Complete the installation of the crankshaft in reverse order of removal. Be sure that the timing marks on the crankshaft gear and camshaft gear are aligned. See Fig. FO36.

CRANKSHAFT OIL SEALS

43. The crankshaft front and rear oil seals are two-piece, molded packing ring type. The lower half of the seals is located in the oil pan. The upper half of the front seal is fitted in a groove in the cylinder block. The upper half of the rear seal is located in a grooved retainer attached to rear of engine block. The seals can be renewed without removing the crankshaft, if desired, after the oil pan is removed as outlined in paragraph 45. Push the old seal halves out of their grooves. Soak new seals in engine oil for two hours before installation. Work the new seal halves into their retaining grooves.

FLYWHEEL

44. REMOVE AND REINSTALL. The flywheel can be removed after splitting the tractor between the engine and transmission housing as outlined in paragraph 21. Unbolt and remove the clutch assembly and the flywheel. It may be necessary to tap the flywheel with a soft face hammer to loosen it from the two dowels used to locate the flywheel on the crankshaft flange.

The clutch friction surface of the flywheel may be resurfaced if it is scored or grooved, providing the thickness of the flywheel, measured between friction surface and mounting flange surface is not reduced to less than 0.855 inch (21.7 mm). Inspect the ring gear for cracks, broken or chipped teeth and excessively worn teeth. To remove the ring gear, drill a hole partially through the ring gear at the root of two teeth. Then use a hammer and cold chisel to crack the ring gear at the drilled location and drive the ring gear off the flywheel. Heat the replacement ring gear to approximately 350° F (175° C). Use caution not to overheat the ring gear as the hardness of the gear teeth could be affected. Quickly install the heated ring gear on the flywheel, then quench the gear teeth with water to obtain a shrink fit.

> NOTE: Flywheels as installed in production Models 8N prior to Serial Number 263844 and all 9N and 2N tractors are not marked to indicate piston position. Model 8N tractors after Serial Number 263843 (with angle drive distributor) are equipped with flywheels which are stamped with an "O" mark indicating top center position and with additional stamped lines from one to 20°. These marks appear in two places on the flywheel, 180° apart.

To reinstall the flywheel, reverse the removal procedure while noting the following special instructions: Tighten the flywheel retaining cap screws to a torque of 75-85 ft.-lbs. (102-115 N•m). Make certain there is no oil or grease on flywheel friction surface. Use a dummy transmission input shaft to align the clutch disc hub with the flywheel pilot bearing when installing the clutch.

OIL PAN

45. REMOVE AND REINSTALL. The cast iron oil pan also forms part of the tractor frame to which the front axle support is bolted. The oil pump screen is a part of the oil pan drain plug and may be cleaned after the plug is removed. The lower halves of the crankshaft front and rear oil seals are located in grooves in the oil pan.

To remove the oil pan from the tractor, first drain the oil. Support the engine with a chain hoist or block up under the transmission housing. Disconnect front axle radius rods and drag links. Unbolt axle support and move the axle assembly forward to clear the oil pan. Remove oil pan mounting cap screws and lower the pan from the engine. Install the oil pan with new gaskets. Tighten the oil pan cap screws to a torque of 15-18 ft.-lbs. (21-24 N•m).

OIL PUMP

46. The oil pump body is a part of the front main bearing cap and can be removed after the oil pan is removed (Fig. FO46). The pump is a gear type and is driven by the crankshaft timing gear. Pump pressure is regulated by a spring-loaded, plunger type bypass valve located on top left side of the timing gear cover as shown in Fig. FO47.

If driven gears in pump body are worn, pitted or chipped, they should be renewed. Recommended clearance of gear shaft in bushing is 0.0005-0.0015 inch (0.013-0.038 mm). Renew fiber drive gear and shaft assembly if gear teeth are worn or damaged, or if shaft journal surface is roughened or worn to less than 0.560 inch (14.22 mm) diameter. The drive gear shaft bushing should be renewed if worn to larger than 0.566 inch (14.38 mm) diameter. New bushing should be sized after installation to provide desired 0.0005-0.0015 inch (0.013-0.038 mm) clearance on a new shaft.

Refer to the following specifications:

Relief Valve Spring Tension . . 31-33 oz. @ 1.380 in.
                                 (8.6-9.1 N @ 35 mm)
Gearshaft Clearance in Bushing. . 0.0005-0.0015 in.
                                 (0.013-0.038 mm)
Idler Gear Clearance on Shaft . . . 0.0025-0.0045 in.
                                 (0.064-0.114 mm)
Gear Backlash . . . . 0.003-0.004 in.
                                 (0.08-0.10 mm)

Install oil pump, main bearing cap and bearing insert. Tighten bearing cap retaining cap screws to a torque of 90-100 ft.-lbs. (122-135 N·m).

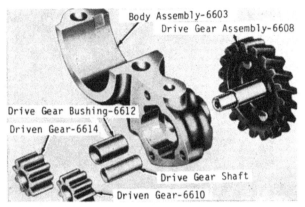

Fig. FO46—Disassembled view of engine oil pump used on Models 9N, 2N, and 8N.

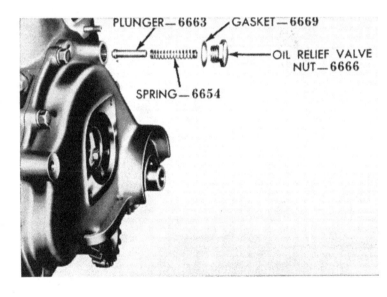

Fig. FO47—Disassembled view of oil pressure relief valve mounted on the front of timing gear cover on Models 9N, 2N and 8N.

FUEL SYSTEM

47. Original carburetor equipment used on all tractors was Marvel-Schebler TSX Series updraft type with an air-bleed compensated main jet and a back suction economizer system. See Fig. FO48 for a cutaway view of a typical carburetor and to the following for carburetor models.

All 9N and 2N tractors TSX33
8N Models to Serial No. 313112 . . . . TSX241
8N Models after Serial No. 313112 . . . . . TSX241B

On all models, a removable fuel strainer is located in the carburetor fuel inlet elbow. The screen should

be removed and cleaned if there is an indication that the engine is starving for fuel. Make sure that the packing located in the drain hole at the bottom of the carburetor air intake is in place or dirt and dust will be drawn into the engine.

CARBURETOR OPERATION

48. IDLE SYSTEM. Idling system fuel is drawn through idle jet (6—Fig. FO49) from main jet well through passage (13). Air is admitted through pas-

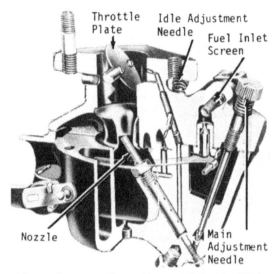

Fig. FO48—Cutaway view of Marvel-Schebler TSX series carburetor used on all models.

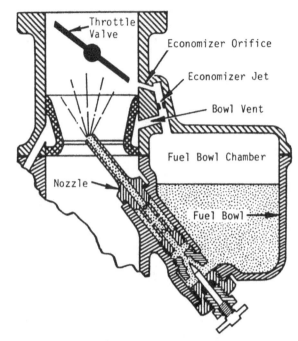

Fig. FO50—Marvel-Schebler economizer system.

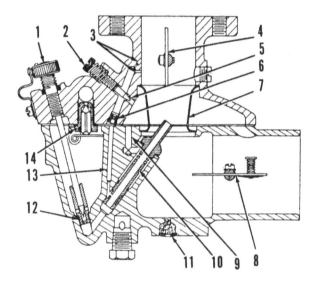

Fig. FO49—Sectional view of TSX series carburetor.

1. Power adjusting needle	8. Choke plate
2. Idle adjusting needle	9. Nozzle air bleed passage
3. Idle ports	10. Discharge nozzle
4. Throttle plate	11. Air horn drain
5. Idle air passage	12. Power jet
6. Idle jet	13. Idle fuel passage
7. Venturi	14. Inlet needle & seat

sage (5) which opens behind the venturi (7). The amount of air admitted and mixed with the fuel is regulated by idle adjusting needle (2). Turning needle in towards its seat reduces the volume of air and enriches the mixture which is discharged at idle ports (3).

49. POWER RANGE. Discharge nozzle (10—Fig. FO49) receives fuel from the float chamber through a passage below the power jet (12). Fuel volume is controlled by power adjusting needle (1) which reduces fuel flow when turned in toward its seat. Air is

bled into the main nozzle through passage (9) and holes drilled in the side of the nozzle.

50. ECONOMY RANGE. The back suction economizer system consists of a sealed float bowl and a calibrated economizer orifice which opens the bowl chamber to varying pressures according to the position of the throttle with respect to the economizer orifice. Air enters the fuel bowl chamber (Fig. FO50) through the air cleaner and the bowl vent. The amount of air drawn out of the chamber is controlled by the economizer jet and the position of the throttle valve.

As the throttle valve is moved from closed towards open position, the economizer orifice is gradually exposed to manifold suction and air flows from the bowl chamber through the jet and out the economizer orifice. This loss of air out through the economizer orifice lowers the pressure (creates a vacuum) in the bowl chamber because air cannot enter through the bowl vent as fast as it leaves through the economizer.

As a result of the lowered pressure (vacuum) in the bowl chamber, fuel flow to the discharge nozzle will be retarded and a leaner or economy mixture will be delivered to the engine. However as the throttle valve nears the wide open position, the suction (vacuum) at the economizer orifice and in the fuel bowl chamber decreases, which causes a faster flow of fuel from the bowl to provide the richer mixture needed for full power.

TROUBLESHOOTING

51. Fuel system problems must be isolated to the fuel shut-off valve, fuel line or the carburetor. The following procedures assume that the ignition system is working properly and is correctly adjusted.

Engine Will Not Start

a. Make sure that there is gas in the tank and that fuel shut-off valve is open. Disconnect fuel line from carburetor and make certain that fuel is being delivered to the carburetor.

b. Check for water in the fuel tank. Drain any accumulation of water and sediment from fuel tank and fill with fresh gasoline.

c. Remove the hose from air intake on carburetor. Apply choke and crank engine several revolutions. There should be evidence of raw fuel in carburetor air intake. If not, check fuel filter elbow on carburetor, float condition and adjustment, and fuel inlet valve and seat. If necessary, rebuild or renew the carburetor.

Rough Idle Or Engine Miss With Frequent Stalling

a. Check choke linkage for proper adjustment.

b. Check throttle stop screw adjustment and set low idle speed to specifications.

c. Check for air leak in the intake system. Air leaking past a worn throttle shaft or defective carburetor or intake manifold gaskets will upset the air:fuel mixture, resulting in a lean air:fuel mixture.

d. Check float adjustment.

Engine "Diesels" (Continues to Run) When Ignition Is Switched Off

a. Check idle mixture (probably too lean), ignition timing and idle speed (probably too fast).

b. Check for engine overheating.

Engine Misses At High Speed Or Lacks Power

a. This indicates possible fuel starvation. Check for plugged filter screen in fuel inlet elbow in carburetor.

b. Check carburetor high speed mixture setting.

c. Check float setting.

Engine Will Not Reach Top Speed and Power is Reduced

a. Check throttle linkage for binding.

b. Check for incorrect float drop, a clogged fuel filter or faulty governor.

c. Check for clogged air cleaner.

d. Check carburetor high speed mixture setting.

Black Exhaust Smoke

a. Check for an excessively rich carburetor air:fuel mixture setting.

b. Check choke cable setting.

c. Check for plugged air filter.

d. Check for leaky float.

e. Check for foreign material under float needle or worn needle valve.

f. Check float setting.

CARBURETOR OVERHAUL

52. R&R AND OVERHAUL. To remove the carburetor, shut off the fuel and disconnect the fuel line. Disconnect the choke and governor control rods. Disconnect the air cleaner hose. Unbolt and remove the carburetor from the manifold.

To disassemble, remove the main adjustment needle (5—Fig. FO51) and idle adjustment needle (4). Remove the screws attaching the fuel bowl (19) to the throttle body (10). Remove float pin (17) and separate the float (16) and float needle valve (14) from throttle body. Unscrew and remove float needle seat and gasket (13). Remove throttle plate retaining screws and remove throttle plate (1) and shaft (2) from throttle body. Pull throttle shaft packing (9) from body. Remove choke plate retaining screws and remove choke plate (25) and shaft (22) from float bowl. Pull choke shaft packing (24) from float bowl bore.

Thoroughly clean all passages in the carburetor using a suitable solvent and compressed air. Position the throttle body and float assembly upside down so the float seats the fuel inlet needle on its seat. The float height should be ¼ inch (6.35 mm), measured from the nearest face of the float to the gasket surface of the throttle body (Fig. FO52). Bend the float lever, if necessary, to obtain the recommended setting.

A carburetor repair kit that contains components necessary for servicing the carburetor is available from Ford. To reassemble carburetor, reverse disassembly procedure using new gaskets and packing. Initial setting of idle mixture adjustment needle (4—Fig. FO51) and main fuel mixture adjustment needle (5) is one turn open from a lightly seated position.

Install carburetor on intake manifold using new mounting gasket. Connect fuel line, choke and throttle control rods and air cleaner hose.

53. ADJUSTMENT. After carburetor is reassembled and installed, run the engine until normal operating temperature is reached. Move the engine speed control lever to closed (idle) position. Adjust the throttle lever stop screw (1—Fig. FO53) on carburetor throttle lever to set idle speed at 400 rpm.

The idle mixture adjustment needle (2) controls the air, turning the screw in (clockwise) enriches the mixture. Initial needle setting is approximately one turn open (counterclockwise) from the seated position. With engine running at idle speed, turn the idle adjusting screw in until the engine falters from an overly rich mixture, then turn the screw out until the engine runs smoothly.

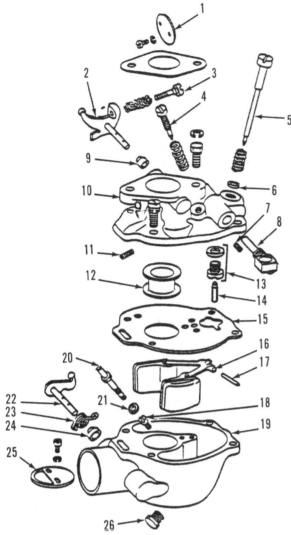

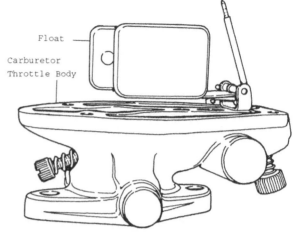

Fig. FO52—Float height should be 1/4 inch (6.35 mm), measured from the nearest face of the float to the gasket surface of the throttle body.

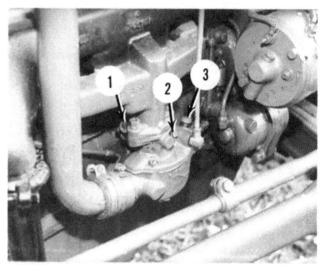

Fig. FO51—Exploded view of Marvel-Schebler carburetor assembly.

1. Throttle plate	14. Float needle valve
2. Throttle shaft	15. Gasket
3. Idle speed adjustment screw	16. Float
4. Idle mixture adjustment needle	17. Pin
5. Main adjustment needle	18. Main jet
6. Gasket	19. Carburetor body
7. Economizer jet	20. Main nozzle
8. Fuel strainer & elbow	21. Gasket
9. Packing	22. Choke shaft
10. Carburetor upper body	23. Choke return spring
11. Idle jet	24. Packing
12. Venturi	25. Choke plate
13. Fuel inlet seat & gasket	26. Drain plug

Fig. FO53—Carburetor adjustment points on Model 9N. Other models are similar.

1. Idle speed adjustment screw (back side of carburetor)	2. Idle mixture adjustment needle
	3. Main adjustment needle

The main adjustment needle (3) controls the fuel flow. Turning the needle in (clockwise) leans the mixture. Initial setting for main needle is one turn open (counterclockwise) from the seated position. The final adjustment of the main needle should be made with the engine running at governed speed under full load. Turn the adjustment screw in until the engine just begins to lose power, then turn the screw out until the power picks up and the engine runs smoothly. If the engine hesitates or backfires when picking up a load, turn adjusting screw out about 1/8 turn more to enrich the mixture.

GOVERNOR

55. The governor used on Models 9N, 2N and 8N is a variable speed, centrifugal ball type, driven by the engine timing gears and interconnected with the hand throttle linkage and the carburetor throttle plate. The governor changes the carburetor throttle setting as required to maintain a given engine speed under varying engine loads.

Splash lubrication for the governor unit is provided from the timing gear case (Fig. FO54), and it is important that the large oil passage opening in the governor ball bearing carrier plate be installed at the top (or up) position to ensure proper lubrication. Further lubrication is provided by an oil line which connects between rear of the governor housing and the oil filter as shown in Fig. FO55. Early model tractors were not provided with the oil line, but service parts are available to permit this modernization.

TROUBLESHOOTING

56. The following are some governor related problems that may occur during operation and their possible causes. Note that external adjustments are responsible for most governor problems.

Hand Throttle Inoperative On First Section Of Quadrant. This is caused by end play in governor spring. The spring should be adjusted as outlined in paragraph 57 or renewed if it is stretched or has lost tension.

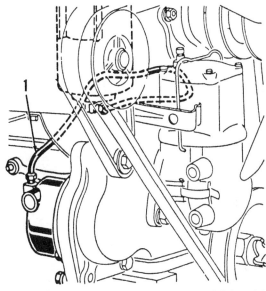

Fig. FO55—On late production tractors, additional lubrication for the governor is provided by an external oil line (1) which connects between rear of governor housing and oil filter.

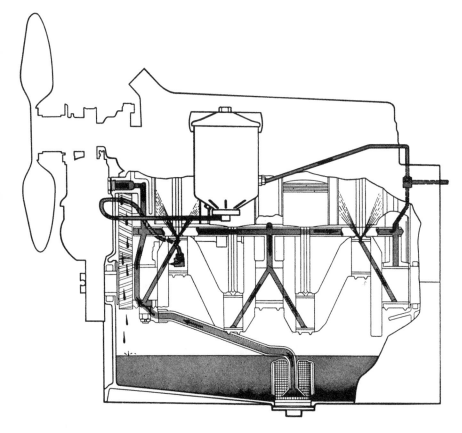

Fig. FO54—Lubricating oil circuit of 9N-2N-8N engine showing path of pressure oil and lubrication of walls and gears.

Fig. FO56—Model 8N governor. Models 9N and 2N are similar.

1. Inner lever
2. Spring
3. Outer lever
4. Oil line
5. Stop screw
6. Governor housing
9. Proofmeter cable

Engine Idle Speed Too High. This may be caused by preload on governor spring. The spring should be adjusted as outlined in paragraph 57.

Engine Will Not Reach Specified Top Speed. This may be caused by incorrect adjustment of either the governor control rod or the governor speed adjust-

ment stop screw. Adjustment procedure is outlined in paragraph 57.

Engine Speed Control Is Erratic Or Sluggish. This may be caused by binding in the governor or governor linkage. The governor-to-carburetor rod must be straight and the ball joints and throttle shaft must move freely. Governor internal components may be worn or binding.

57. ADJUSTMENT. The governor spring (2—Fig. FO56) should be a snug fit, but should have no end play and no preload. The spring setting can be adjusted by bending the loop in the spring with pliers.

Correct engine governed speed (no-load) with hand throttle in wide open position is 2200 engine rpm, 800 pto rpm or 1494 belt pulley rpm. To adjust maximum speed setting, turn the maximum speed stop screw (5—Fig. FO56) in (clockwise) to reduce rpm, or out (counterclockwise) to increase rpm.

If the governor lever reaches the maximum speed stop setting before the hand throttle is in the last notch on the quadrant (wide open) or if governor lever fails to reach the maximum speed stop with the hand lever in the last quadrant notch, adjust the control linkage as follows: On 9N and 2N models, hold the throttle shaft lever (A—Fig. FO57) and bend the linkage until speed is correct. On 8N models, adjust the length of the governor link rod (Fig. FO58) until the governor lever contacts the maximum speed stop when the throttle lever is in the last notch of the quadrant.

58. R&R AND OVERHAUL. To remove the governor, disconnect the control links from governor levers (1 and 3—Fig. FO56). Disconnect the oil supply line (4) and Proofmeter cable (9) if so equipped. Unbolt and remove the governor assembly.

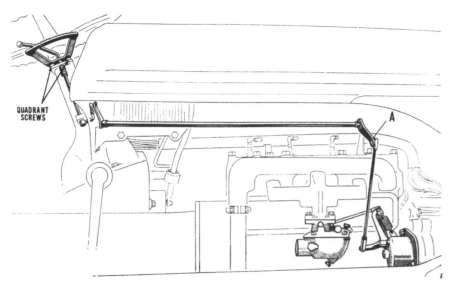

Fig. FO57—On 9N and 2N tractors, hold throttle linkage at point "A" and bend linkage, if necessary, to obtain desired throttle movement.

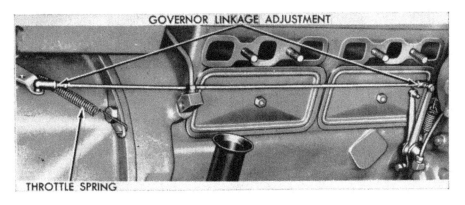

Fig. FO58—On 8N tractors, turn the governor link rod to lengthen or shorten the rod so throttle lever is in the last notch in the quadrant when governor lever contacts the stop.

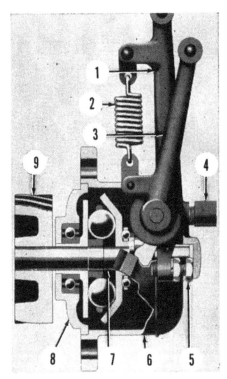

Fig. FO59—Cutaway view of centrifugal ball type governor used on 9N-2N-8N tractors.

1. Inner lever
2. Spring
3. Outer lever
4. Oiler elbow
5. Stop screw

6. Housing
7. Flyball unit
8. Bearing retainer
9. Drive gear

To disassemble, remove the cap screws attaching the ball bearing base (8—Fig. FO59) to the governor housing (6) and withdraw flyball unit (7), drive shaft and driven gear as an assembly from the body. Remove spring clip (18195) and disassemble shaft and flyball components. See Fig. FO60. Drive the tapered pin (Fig. FO61) from the governor fork and control lever shaft. Withdraw control lever assembly from governor housing.

Examine flyballs and renew if they have flat spots, pits or are excessively worn. Inner surface of cone shaped upper race should be smooth and even. If grooved or pitted, renew the race. Check condition of fork base and thrust bearing, drive shaft bearing and driven gear. Renew worn or damaged parts. Reassemble the flyball and drive shaft components, then check the clearance between the washer (356430-S) and fork base (18194). See Fig. FO60. The clearance should be 0.220-0.230 inch (5.59-5.84 mm) and is adjusted by varying the number of shims (18241) until correct clearance is obtained.

If available, governor setting gauge No. EL07691 shown in Fig. FO62 may be used to check the clearance adjustment as follows: Clamp shaft and driver assembly in gauge as shown, and insert the Go-No Go gauge between the washer and fork base. If only thin end of gauge can be inserted, clearance is satisfactory. If thick section of gauge enters the gap, excessive

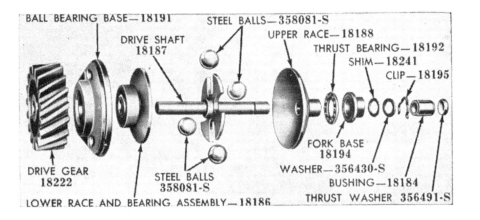

Fig. FO60—Exploded view of engine speed governor flyball assembly used on 9N-2N-8N tractors.

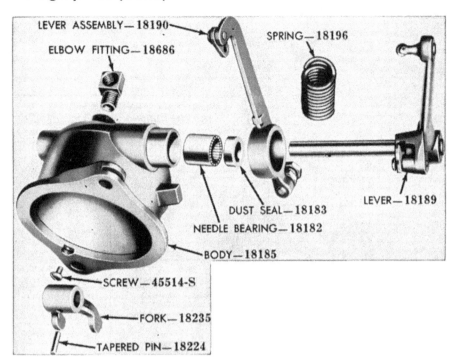

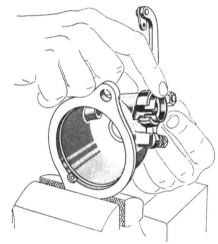

Fig. FO61—Exploded view of governor housing and lever assemblies.

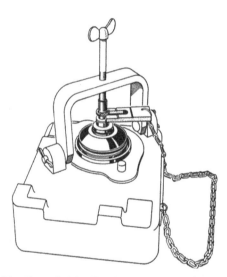

Fig. FO62—If available, Ford governor setting gauge No. ELO7691 can be used to measure clearance between washer and governor fork.

Fig. FO63—If inner lever is loose, a shim may be installed to remove excess clearance.

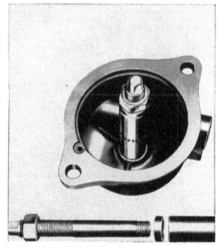

Fig. FO64—Use a suitable tool to remove outer lever shaft bushing from governor housing.

clearance exists and thin shims should be added until clearance is correct.

Check governor lever shaft assembly and needle bearing (Fig. FO61) for binding or excessive looseness and renew parts where required. If inner control lever assembly is loose, renew lever and/or housing or insert a $1/2 \times 2^7/8$ inch (13×73 mm) shim around housing boss as shown in Fig. FO63. Check governor shaft bushing (18184—Fig. FO60) for wear and if necessary, renew the bushing using suitable remov-

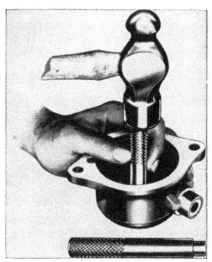

Fig. FO65—Use a piloted bushing driver to install new shaft bushing.

ing and installing tools as shown in Figs. FO64 and FO65.

NOTE: Make certain that the thrust washer (356491-S) shown in Fig. FO60 is in place before installing bushing.

Assemble governor control levers and fork in housing (Fig. FO61). Assemble drive shaft and flyball components and secure with spring clip (Fig. FO60). Install driven gear, ball bearing base, drive shaft and flyball assembly in governor housing (Fig. FO59).

Install governor unit on the engine. Connect Proofmeter cable (if used), oil line and control links. Check governed engine speed and adjust as outlined in paragraph 57 if necessary.

COOLING SYSTEM

RADIATOR

59. REMOVE AND REINSTALL. Shut off the fuel and disconnect the fuel line at the carburetor. Remove the four cap screws that secure the rear of the hood to the instrument panel. Remove the two cap screws that attach front of the hood to the axle support, and remove the radiator grille. Remove the air cleaner grill or precleaner assembly from the side of the hood on 8N tractors. Remove precleaner stack on other models if tractor is so equipped. Unfasten the wiring from the hood. Lift the hood and fuel tank off the tractor. Drain the cooling system and disconnect

the radiator hoses, then unbolt and lift the radiator from the front support. To install the radiator, reverse the removal procedure.

WATER PUMP

All Models

60. REMOVE AND REINSTALL. To remove the water pump (1—Fig. FO66), first drain the cooling system. Remove the fan belt and disconnect the radiator hose from the pump. Unbolt the fan blades and move them forward into fan shroud to provide clearance for pump removal. Remove the cap screws attaching the pump assembly to the cylinder block and withdraw the pump. To reinstall the water pump, reverse the removal procedure.

Adjust the fan belt tension by moving the generator until belt can be deflected ½ inch (13 mm) with moderate thumb pressure, measured at a point midway between generator and crankshaft pulleys.

Models 9N and 2N

61. OVERHAUL. To disassemble the removed pump, press fan pulley (5—Fig. FO67) off shaft and bearing assembly (4) using a suitable puller. Remove cover plate from rear of pump housing. Remove bearing retainer snap ring from front of pump body, then press the shaft and bearing unit forward out of impeller and pump body. Remove snap ring from impeller, then remove composition seal washer, spring retainer and spring from impeller hub.

Inspect for worn or damaged parts. The bushing (3) must be flat over its entire face area and square

Fig. FO66—Fan belt should deflect 1/2 inch (13 mm) with moderate thumb pressure, measured at a point midway between generator and crankshaft pulleys. Fan belt tension is adjusted by moving the generator (Model 9N shown).

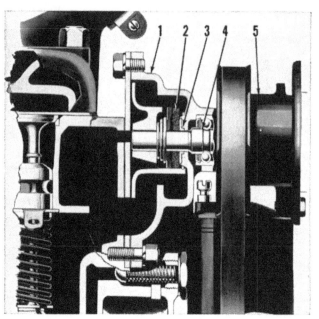

Fig. FO67—Cross-sectional view of water pump used on 9N and 2N tractors.

1. Pump housing
2. Seal assy.
3. Bushing (seal seat)
4. Bearing & shaft
5. Pulley

within 0.001 inch (0.025 mm) to provide a satisfactory sealing surface. Bushing face may be resurfaced if not excessively worn. A pump repair kit, consisting of seal assembly and gaskets, and a complete water pump assembly are available from Ford.

To reassemble the water pump, press the shaft and bearing assembly into the pump housing until bearing is bottomed against housing counterbore. Press

only against outer race of the bearing. Assemble bushing (3) and seal assembly on the shaft. Support front end of the shaft, then press the impeller onto rear of shaft until rear face of impeller is approximately 1/32 inch (0.8 mm) below rear surface of the pump housing. Support the rear end of the shaft and press the fan pulley onto front of shaft. Install rear cover plate using a new gasket.

Model 8N

62. OVERHAUL. To disassemble the water pump, remove the lock ring (1—Fig. FO68) and rear cover plate. Pull the fan pulley (2) off the pump shaft using a suitable puller. Extract bearing retaining ring (4), then press shaft and bearing (3), seal (6) and impeller (7) as a unit rearward out of pump housing. Press the impeller off the shaft and remove the seal assembly (6) and slinger washer (5).

Inspect for worn or damaged parts. A pump repair kit, consisting of seal assembly and gaskets, and a complete water pump assembly are available from Ford.

To reassemble the pump, press shaft and bearing assembly into the pump housing from the rear using a suitable size sleeve to apply pressure only against the outer race of the bearing. Do not press against the end of the shaft as damage to the bearing could result. Install the bearing retaining ring (4) into groove in the bearing outer race. Assemble the slinger washer (5) and seal assembly (6) on the shaft. Support the front of the shaft, then press the impeller onto the shaft until rear face of impeller is approximately 1/32 inch (0.8 mm) below the rear face of the housing.

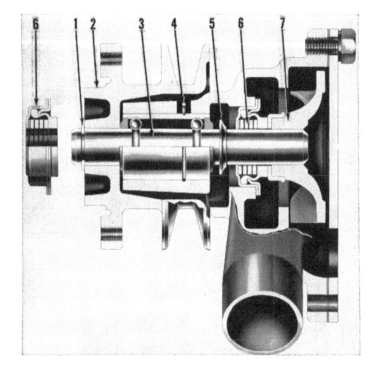

Fig. FO68—Cross-sectional view of water pump used on 8N tractors.

1. Lock ring
2. Pulley
3. Bearing & shaft
4. Retaining ring
5. Slinger
6. Seal assy.
7. Impeller

Support rear of the shaft, then press the pulley onto front of shaft until the lock ring (1) can be installed. Install the rear cover plate with a new gasket.

THERMOSTAT

63. A thermostat is located in the upper radiator hose. The thermostat should start to open at 160-165° F (71-74° C) and be fully open at 190-200° F (88-93° C).

When installing the thermostat into the radiator hose, be sure the bi-metal spiral end is toward the engine. See Fig. FO69.

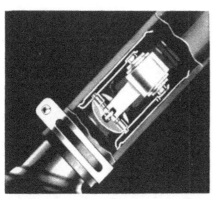

Fig. FO69—Cooling system thermostat used on 9N-2N-8N models is located in the upper radiator hose.

ELECTRICAL AND IGNITION SYSTEM

CHARGING SYSTEM

64. TROUBLESHOOTING. The 9N, 2N and 8N tractors were orginally equipped with a 6-volt, positive ground electrical system. Refer to Fig. FO70A, FO70B, Fig. FO71 or Fig FO72 for tractor wiring diagrams.

Charging system troubles are generally caused by a defective generator, voltage regulator or battery. Many electrical problems can be traced to a simple cause such as a blown fuse, a loose or corroded connection, a loose generator drive belt or a frayed wire. The following are symptoms of problems you may encounter.

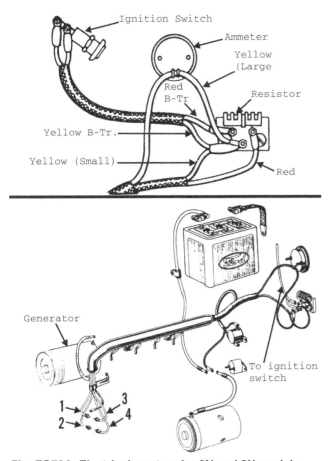

Fig. FO70A–Electrical system for 9N and 2N models.

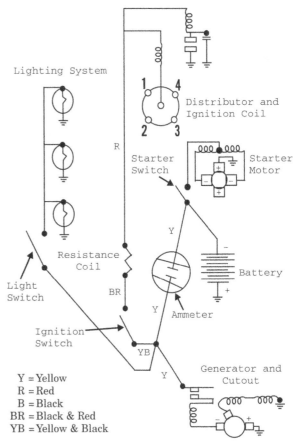

Y = Yellow
R = Red
B = Black
BR = Black & Red
YB = Yellow & Black

Fig. FO70B–Schematic wiring diagram for 9N and 2N models.

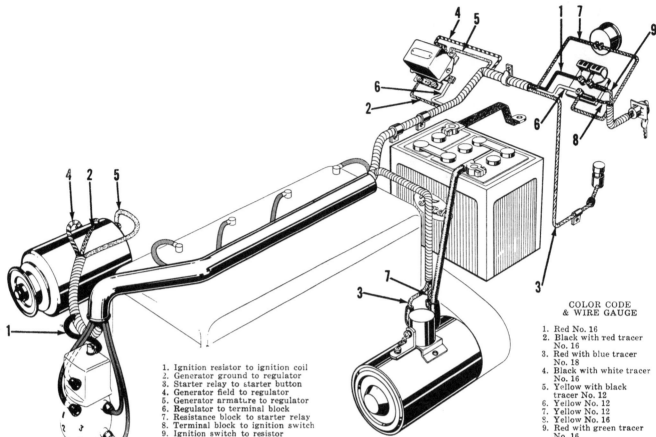

1. Ignition resistor to ignition coil
2. Generator ground to regulator
3. Starter relay to starter button
4. Generator field to regulator
5. Generator armature to regulator
6. Regulator to terminal block
7. Resistance block to starter relay
8. Terminal block to ignition switch
9. Ignition switch to resistor

COLOR CODE
& WIRE GAUGE

1. Red No. 16
2. Black with red tracer No. 16
3. Red with blue tracer No. 18
4. Black with white tracer No. 16
5. Yellow with black tracer No. 12
6. Yellow No. 12
7. Yellow No. 12
8. Yellow No. 16
9. Red with green tracer No. 16

Fig. FO71—Electrical system on Model 8N prior to serial number 263844. Wire color codes and circuits are as follows:

1. Red—Ignition resistor to ignition coil
2. Black with red tracer—Generator ground to regulator
3. Red with blue tracer—Starter relay to starter button
4. Black with white tracer—Generator field to regulator
5. Yellow with black tracer—Generator armature to regulator
6. Yellow—Regulator to terminal block
7. Yellow—Resistance block to starter relay
8. Yellow—Terminal block to ignition switch
9. Red with green tracer—Ignition switch to resistor

Battery Frequently Discharged - Ammeter Indicates No Discharge. This can be caused by a drive belt that is slightly loose. With the engine OFF, grasp the generator pulley with both hands and try to turn it. If the pulley can be turned without moving the belt, the drive belt is too loose. As a rule, keep the belt tight enough so that it can be deflected only about ½ inch (13 mm) under moderate thumb pressure applied between the pulleys. The battery may also be at fault.

As a quick check for low generator output, connect a jumper wire between the armature and field terminals on the generator. Start engine and note ammeter reading, then remove jumper wire and again note ammeter reading. If charging rate increased when jumper wire was disconnected, generator is operating satisfactorily; check for faulty voltage regulator. If voltage regulator is operating satisfactorily, check for poor wiring connections causing high resistance. If there was no change in charging rate when jumper wire was disconnected, check for faulty generator and repair or renew.

Ammeter Needle Fluctuates Between Charge And Discharge. This usually indicates that the charging system is working intermittently. Check drive belt tension first, then check all electrical connections in the charging circuit. As a last resort, check the generator and regulator.

Battery Requires Frequent Addition Of Water. The generator is probably overcharging the battery. Check and if necessary, renew the voltage regulator.

GENERATOR AND REGULATOR

65. A two-brush, shunt wound, 6-volt generator, part No. 8N10000B, provided with an external vibrating type voltage regulator replaces the early three-brush unit, part No. 9N10000C, 2N 10000 and 8N10000A units used on 9N, 2N and early 8N tractors. The generator used on 8N tractors after Serial No. 8N263843 is part No. 8N10001.

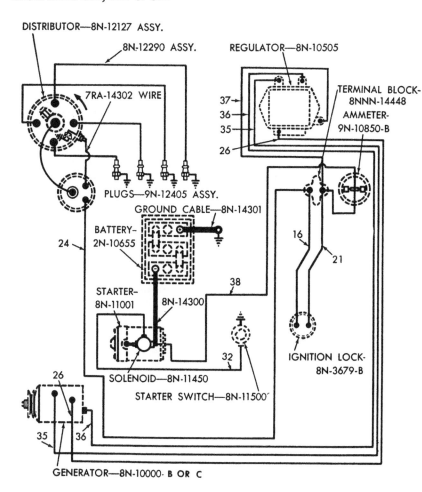

Fig. FO72—Schematic wiring diagram on 8N tractors (with two brush generators and angle drive distributors) after serial number 263844. Wire color codes and circuits are as follows:

16. Red with green tracer—Terminal block to ignition lock
21. Yellow—Terminal block to ignition lock
24. Red—Coil to terminal block
26. Black with red tracer—Generator ground to regulator
32. Red with blue tracer—Starter solenoid to starter switch
35. Black with white tracer—Generator field to regulator
36. Yellow with black tracer—Generator armature to regulator
37. Yellow—Regulator to terminal block
38. Yellow—Starter solenoid to terminal block

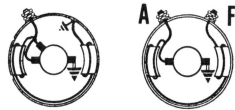

Fig. FO73—Internal circuits of a third brush (left view) and a two brush shunt type (right view) generator.

The shunt wound generator produces a higher output at both high and low speeds than the third brush generator. The internal wiring circuits of both types of generators are shown in Fig. FO73.

Refer to the following specifications:

Generator No. 2N10000, 9N10000C
Field Resistance @ 70° F . . . . 1.5 Ohms
Brush Length (minimum) . . . . 0.350 in.
(8.89 mm)
Output (rated)
　Maximum Amperes . . . . . 11.5 @ 925 engine rpm
　Volts 6.5

Generator No. 8N10000, 8N10000A
Field Resistance @ 70° F . . . . 4.0 Ohms
Brush length (minimum) . . . . 0.350 in.
(8.89 mm)
Output (rated)
　Maximum Amperes . . . . 11.5 @ 1500 engine rpm
　Volts 6.5
Regulator
　Cutout Closing Voltage . . . . 6.0-6.5 volts
　Voltage Limiter . . . . 6.9-7.4 volts
　Cutout Opening Voltage . . 0.5-1.5 volts less than
closing voltage or 6 amps
maximum reverse current
@ 6.0 volts minimum

Generator No. 8N10000B, 8N10001
Field Resistance @ 70° F . . . . 3.2 Ohms
Brush Length (minimum) . . . . 0.400 in.
(10.16 mm)
Output (rated)
　Maximum Amperes . . . . . . 20 @ 1650 engine rpm
　Volts 7.0
Regulator
　Cutout Closing Voltage . . . . 6.0-6.6 volts
　Voltage Limiter . . . . 7.1-7.5 volts

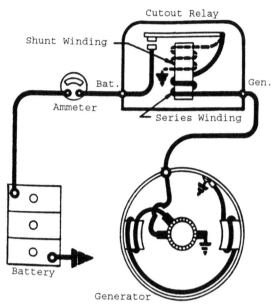

Fig. FO74—Schematic drawing of typical charging system with cutout relay.

Cutout Opening Voltage .. 0.5-1.5 volts less than closing voltage or 6 amps maximum reverse current @ 6.0 volts minimum

Cutout Relay

66. A typical cutout relay is shown schematically in Fig. FO74. The cutout relay is an electromagnetic switch which opens and closes the generator-to-battery circuit. When generator operating voltage is higher than battery voltage, the cutout relay contacts close and current passes from generator to battery. When the generator slows or stops, the cutout relay contacts open and break the circuit to prevent the battery from discharging back through the generator.

The cutout relay unit 9N10505B used with 3 brush generators should be adjusted to a closing voltage of 6.5-7.25 volts. Opening voltage should be ½ to 1½ volts less than the closing voltage or 6 amperes maximum reverse current at 6 volts minimum.

To check and adjust closing voltage, connect a voltmeter between the "GEN" terminal of relay and the ground. Slowly increase generator speed and note the voltage at which the contacts close. Closing voltage is adjusted by changing the spring tension on the relay armature. Increase spring tension to raise closing voltage and decrease spring tension to lower closing voltage.

Opening voltage or amperage is adjusted by changing the air gap between the stationary contact point and the movable armature contact point.

Voltage Regulator

67. The combined cutout relay and voltage regulator assemblies 8N10505, 8N10505A and 8N10505B used on 2 brush mounted generators have a cutout unit (points normally open) and a voltage regulator unit (fine winding, points normally closed) on the same base. The cutout relay unit of this assembly should be adjusted to the specifications listed in paragraph 66. Refer to paragraph 66 for adjustment procedure.

The vibrating voltage regulating unit settings should be 6.9-7.4 volts for generators 8N10000, 8N10000A or 7.1-7.5 volts for generators 8N10000B, 8N1000l with the generator at normal operating temperature and generator current of 5 amperes. With generator maximum current of 20 amperes, the voltage should not exceed 6.8 volts. To check and adjust voltage setting, connect voltmeter to the regulator "BAT" terminal and to regulator base. Connect ammeter in series with "BAT" terminal and battery lead wire. Operate generator and check voltage reading. Adjustment is made by increasing or decreasing the armature spring tension.

Shunt Wound Type Generator

68. Field current strength of shunt type generator is controlled externally by an automatic control unit known as a generator regulator. A regulator consists of a pair of contact points mounted on a movable armature, connected in series with the generator field circuit and a resistance unit (resistor), connected in shunt (parallel) to the contact points and to an electromagnet consisting of an iron core and coil assembly. The contact points are held closed by spring tension and are opened by generator current, which induces a magnetic pull of sufficient strength in the iron core of the regulator electromagnet to overcome the spring tension.

When the points are closed, the field current flows through the contact points (normally closed), completing the field circuit. When the points are opened due to magnetic pull overcoming the spring tension, the resistor is inserted in the field circuit, thus reducing the generator output. This action, however, results in weakening the magnetic field of the regulator electromagnet, permitting spring tension to close the contact points, thus cutting out or bypassing the resistor and completing the cycle. This cycle of regulator action (rapid opening of contacts) is completed many times a second, thus controlling the generator output to meet the varying requirements of the electrical system.

69. QUICK CHECKS (Shunt Wound Systems). If the generator or regulator is not operating properly,

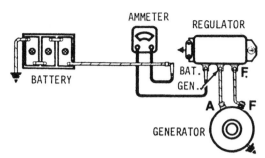

Fig. FO75—Connect a test ammeter in series at the "BAT" terminal of the regulator as shown to check operation of shunt type generator.

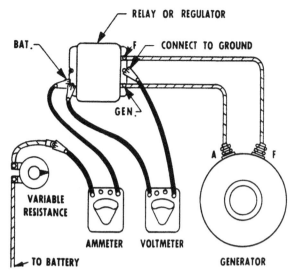

Fig. FO76—Meter connections to check third brush generator output.

the simple tests outlined below can be used to determine which unit is at fault.

A FULLY CHARGED BATTERY AND A HIGH CHARGING RATE indicate improper operation of the equipment and will result in battery overcharge and high system voltage. To check, connect test ammeter into the circuit at the "BAT" terminal of the regulator as shown in Fig. FO75 and disconnect the "F" terminal of the regulator. If the output drops off as the "F" terminal lead is disconnected, check the regulator for a high voltage setting or for grounds or shorts in the regulator. If the output does not drop off under these conditions, then the generator field circuit is grounded, either in the wiring or in the generator itself, and it must be removed for additional checking.

WITH A LOW BATTERY AND A LOW OR ABSENT CHARGING RATE, check the circuit for loose connections or defective leads, since these produce high resistance which will cause the voltage regulator to operate as though the battery were fully charged, even though the battery is in a discharged condition. Next, with test ammeter in charging circuit as in Fig. FO75, operate generator at medium speed and momentarily ground the "F" terminal of the regulator. If the generator output now comes up, check for a low voltage setting or for dirty or oxidized regulator contact points. If the output remains low, remove the generator for further checking. If no output at all is obtained, it might be well to check the cutout relay, since it might not be closing due to a high voltage setting or an open shunt winding.

NOTE: Burned resistance units, windings, or fused regulator contacts result from open circuit operation or high resistance in the charging circuit. Check for these conditions carefully.

Third Brush Type Generator

70. The third brush type generator uses the third brush to provide an internal means of controlling the maximum generator output. The third brush conducts current to field windings from armature commutator. The main brushes contact commutator at two places where the maximum voltage is produced, and the third brush is placed in some position between the main brushes.

By moving the third brush toward the adjacent main brush (with armature rotation), voltage across the field and current through the field windings are raised. This action increases strength of the magnetic field, resulting in higher generator output. Inversely, by moving third brush away from the adjacent main brush (against armature rotation), output will be reduced.

71. CHECKING THIRD BRUSH SYSTEM. Connect an ammeter and a $\frac{1}{4}$ ohm variable resistor in series in the charging circuit between the battery and "BAT" terminal of the cutout relay (Fig. FO76). Connect a voltmeter between the "BAT" terminal and ground. Run the engine at 925 rpm and adjust the variable resistor to obtain 6.5 volts while noting the generator output. Ammeter reading should be 11.5 amperes. The generator output may be increased by moving the third brush in direction of armature rotation, or decreased by moving the third brush against armature rotation. If correct output cannot be obtained by adjusting the third brush, refer to troubleshooting section in paragraph 64.

72. GENERATOR MOTORIZING TEST. If the preceding tests indicate generator trouble, remove the generator from the tractor and place on test bench for motorizing test. A 0-20 ampere ammeter and a fully charged battery are needed to perform the motorizing test.

Connect the ammeter (4—Fig. FO77) in series between the generator "A" terminal and the battery negative lead wire. Connect the battery positive lead wire to the generator frame (2). On generators having

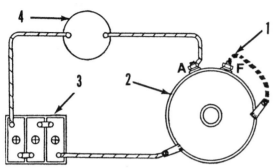

Fig. FO77—Generator motorizing test circuit diagram.

1. Jumper lead   3. Battery
2. Generator   4. Ammeter

two terminals, use a jumper lead (1) to connect the generator "F" terminal to the generator frame as shown in Fig. FO77. With generator connected as described, one of the following conditions should develop.

DOES NOT MOTOR, and ammeter goes off scale. This generally indicates a ground in the main terminal, terminal to brush lead, or in insulated brush holder.

DOES NOT MOTOR, and ammeter reads low. Dirty commutator, oily, worn or stuck brushes, or weak brush springs are the common causes of this condition.

MOTORS SLOWLY AND UNEVENLY with ammeter reading high. A partial short in the generator terminal, lead or brush ring, carbon or copper dust, grounded or shorted armature may cause this trouble. Turn generator pulley by hand, opposite to normal rotation, very slowly. If at any point, a loss of torque occurs and an increase in the current is indicated on ammeter, a grounded or shorted armature coil will usually be found at that point.

MOTORS SLOWLY AND UNEVENLY with ammeter reading low. This lack of motoring is quite commonly due to dirty, rough, oily, or burned commutator, short or dirty brushes, or weak brush springs, or an armature with open or partially open coils. Turn pulley against normal rotation and if at any point torque drops off noticeable and current drops, the cause is probably open armature coils. Brushes should be held down by finger pressure to make sure that they are making contact.

MOTORS SMOOTHLY, but ammeter reads high (above 5-8 amperes). This may be due to high brush spring pressure, binding of dry bearings, bent armature shaft, or improperly set brush ring.

MOTORS SMOOTHLY, but refuses to charge when installed on the tractor. Polarize field poles by momentarily contacting generator field terminal with battery lead, to eliminate the possibility that field polarity was reversed during test.

Clean commutator with 00 sandpaper and "snap" the brushes to be sure that they make good contact.

Vibration may cause an open, short or grounded circuit in armature or field that did not appear until generator was subjected to the vibration incidental to tractor operation.

MOTORS AT EXTREME SPEED due to an open field circuit. A field fuse may be blown, or in case of third brush generators, the third brush may not be making good contact. It is also possible that there is a break in a field winding or connection.

GENERATOR OVERHAUL

73. DISASSEMBLY. Before disassembling the generator, remove the cover band and inspect for thrown solder, which is evidence that the generator has been overheated, caused by an electrical overload or armature drag.

Generator overloading may be caused by: High regulator setting; defective regulator; open charging circuit (some installations); or on third brush generators, the control brush may be too far advanced.

If overheating is due to regulator setting being too high, check and readjust to specifications; if regulator is defective, renew the regulator unit.

Refer to Fig. FO78. Disassemble the generator by first disconnecting the main brush and field leads. Remove the through-bolts and withdraw commutator end frame. Remove the drive end frame and armature from the field frame. Remove pulley and drive end frame from armature. It is not necessary to remove the field coils from field frame unless they are shorted, grounded, open or have charred or otherwise damaged insulation.

The armature and field coils should not be cleaned by use of a degreasing solution or by any high temperature method, as insulation might be damaged. Ball bearings may be packed with a good, nonfiber, high temperature grease unless they are sealed. Internal generator connections should be inspected and resoldered, if necessary, using resin flux solder.

74. INSPECTION AND TESTING. Data on testing and overhauling generator components is outlined in paragraphs 75 through 78.

75. BRUSHES AND HOLDERS. Brushes should be free in their holders and should rest on the commutator with sufficient pressure to assure good contact. If the brushes are worn down more than half their original length, they should be renewed.

Brush holders and brush arms should be firmly attached to their mounting. Inspect the insulated brush holders and arm mounts for broken or charred insulation, and renew if found defective.

Insulated brush holders or brush arms are tested for grounds by using a battery powered test lamp. Place one test lead on the insulated brush holder or arm and place the other test lead on the end frame.

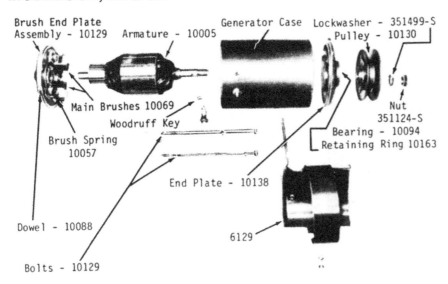

Brush End Plate
Assembly - 10129   Armature - 10005   Generator Case   Lockwasher - 351499-S
                                                          Pulley - 10130

Main Brushes 10069

Woodruff Key

Brush Spring
10057

                                                          Nut
                                                          351124-S

End Plate - 10138

                                    Bearing - 10094
                                    Retaining Ring 10163

Dowel - 10088

6129

Bolts - 10129

Fig. FO78—Disassembled view of generator used on 8N tractors.

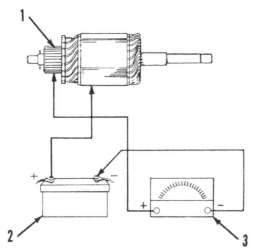

Fig. FO79—Testing generator armature for grounded circuits. A battery powered test lamp can be used in place of the voltmeter (3) if desired.

1. Commutator
2. Battery
3. Voltmeter

If the test lamp lights, the brush holder or arm is grounded and should be renewed.

76. ARMATURE. Visually inspect the armature for mechanical defects such as lamination drag or lamination loose on shaft; worn commutator, burned or high mica, bent shaft or worn bearing surfaces. Make sure that windings are tight in slots and properly soldered to the commutator riser bars. Resolder any loose commutator connections. If the commutator surface is rough, out-of-round, burned, or has high mica, it should be turned down on a lathe and the mica undercut to a depth of $\frac{1}{32}$ inch (0.8 mm). Carefully clean the slots to remove any metal cuttings, then sand lightly with No. 00 sandpaper to remove any copper burrs.

The armature is tested for grounds by using a battery and voltmeter (Fig. FO79) or a battery pow-

ered test lamp. Place one test point on core or shaft (not on bearing surface) and touch the other test point to each commutator segment in turn. If voltage is indicated or the lamp lights, the armature is grounded. Discard a grounded armature if one is found. See Fig. FO79.

The armature is tested for shorts by use of a growler which operates on AC current. Place armature on the growler, close growler circuit, and with a steel strap such as a hack saw blade held slightly above the core, slowly rotate the armature. If the blade or steel strap vibrates, the armature is short circuited and should be renewed.

The armature can be tested for open circuits by using a growler or ohmmeter. Place armature in growler and close growler circuit. Adjust contact fingers so they contact adjacent segments at side of commutator. Observe meter reading and rotate armature to test each succeeding pair of segments. Each should have approximately the same meter reading. A zero reading indicates an open circuit in the winding.

77. FIELD COIL. The field frame and field coils should be cleaned using cloth dampened in cleaning solvent.

CAUTION: Do not soak or submerge coils in solvent.

Field coils are tested for open circuits by using a battery powered test lamp. Place one test point on the field terminal "F" and the other test point on the field lead. If the lamp lights, the field coils are NOT open. If the test lamp does not light, the field circuit is open. Locate open coil by testing each coil for continuous circuit. Renew defective coils.

Field coils are tested for grounded circuit by using the test lamp. If the field circuit is internally

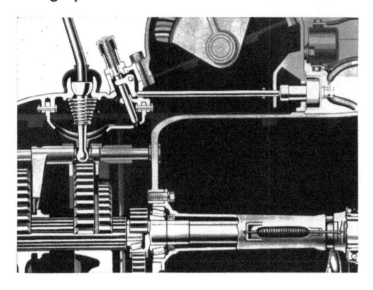

Fig. FO80—Starter switch mechanism used on Models 9N and 2N. A basically similar arrangement is used on Model 8N except that mechanically actuated relay switch is not used.

grounded to the generator frame, disconnect at that point before making the test. Place one test point on the field terminal "F" and the other test point on the generator frame. If the test lamp lights, field coils are grounded. Renew grounded coils.

78. BEARINGS. Check both of the end frame assembly bearings for wear by placing the armature in a soft jawed vise and placing the end frame in position on the armature. Any noticeable side play indicates worn bearing or worn armature shaft. Inspect parts for wear and renew if necessary. Nonsealed ball bearings should be washed in solvent and repacked with a nonfiber, high temperature grease. Sealed ball bearings should not be washed in solvent as they cannot be relubricated without special equipment. Bronze bushings, which are usually of the oil absorbent type, and their felt wicks should be soaked in engine oil before assembly.

79. REASSEMBLY. Assembly is the reverse of disassembly, plus the following: After generator is assembled, check armature end play, which should be 0.003-0.010 inch (0.08-0.25 mm) on most generators. Correct the end play by using spacing washers, if provided, or by renewal of the ball bearing which takes the thrust. Mount the generator on a test stand and operate to make sure that the output is within specifications. After installing the generator on the tractor and before starting the engine, it should be polarized to ensure that it will charge properly.

79. POLARIZING GENERATOR. Whenever generator has been removed and installed and before it is operated, it should be polarized in order to establish correct polarity with respect to the battery it is to charge.

First make all connections between the battery and generator. To polarize a generator where only a cutout

relay is used in the system, use a jumper lead to make a momentary connection between the battery terminal of cutout relay or battery terminal of starter switch and the "A" (armature) terminal of the generator. When a vibrating external regulator is used in the system, make a momentary connection between "BAT" terminal and "GEN" terminal of regulator. This allows a momentary flow of current through the generator, which will establish correct polarity.

STARTING MOTOR AND SWITCH

The Ford 6-volt starter and Bendix drive assembly, No. 8N11001, replaces the No. 9N1101, 9N1002 and 8N1102 assemblies used prior to 1950 production. The 8N1101 starter should have a current draw of: 46-50 amperes at 6 volts running at no load or 100-150 amperes when cranking a warm engine.

The starter button is provided with an interlock as shown in Fig. FO80 which prevents starting the engine when the tractor is in gear.

81. TROUBLESHOOTING. The starting system consists of the battery, starter motor, starter relay, ignition switch, neutral start interlock in the transmission, and connecting wiring. The transmission shift lever must be in neutral position before the starting circuit will operate. Starting problems are relatively easy to find. In most cases, the trouble is a loose or dirty electrical connection.

Slow Cranking Starter. Connect a fully charged booster battery to starter using jumper cables. Listen to the starter cranking speed as the engine turns over. If the cranking speed sounds normal, check the battery for loose or corroded connections or a low charge. Clean and tighten the connections as required. Recharge the battery if necessary. If cranking speed does not sound normal, clean and tighten all starter relay connections and the battery ground on the frame and/or engine.

Starter Relay Clicks, Starter Does Not Crank Engine. Clean and tighten all starter and starter relay connections. Make sure the terminal eyelets are securely fastened to the wire strands and are not corroded. Remove the battery terminal clamps. Clean the clamps and battery posts. Reinstall the clamps and tighten securely. If the starter does not crank, connect the jumper cables. If the starter still does not crank, renew or overhaul the starter.

Starter Relay Chatters, (No Click) Starter Does Not Crank Engine. Check the wire connection at the starter relay. Clean and tighten if necessary. Check the relay mounting screws for good, tight ground. Place the transmission in neutral. Disconnect the wire at the starter relay. Connect a jumper wire between this relay connector and the battery cable terminal on relay. Connect the jumper cables. Try starting the engine. If engine starts, check the ignition switch and system wiring for an open circuit or loose connection. If the engine does not start, renew the starter relay.

Starter Spins But Does Not Crank Engine. Remove the starter. Check the starter pinion gear. If the teeth are chipped or worn, inspect the flywheel ring gear for the same problem. Renew the starter drive and/or ring gear as required. If the pinion gear is in good condition, disassemble the starter and check the armature shaft for corrosion. If there is no corrosion, the starter drive assembly is slipping. Renew the starter drive using a new or rebuilt unit.

82. STARTER NO-LOAD TEST. With starter motor removed from tractor, connect motor in series with a fully charged battery and an ammeter capable of reading several hundred amperes. If an rpm indicator is available, read the armature rpm also. The following are some common no-load test results.

LOW FREE SPEED WITH LOW TORQUE AND HIGH CURRENT DRAW MEANS: (a) Tight or dirty bearings, bent armature shaft or loose pole shoe screws. (b) Grounded armature or field. Check further by raising grounded brushes and insulating them from commutator and checking with a test lamp between the insulated terminal and frame. If test lamp lights, raise other brushes from commutator and check fields and armature separately to determine whether fields or armature are grounded.

FAILURE TO OPERATE WITH HIGH CURRENT DRAW INDICATES: (a) Direct ground in relay switch, terminal or fields. (b) Frozen shaft bearings.

FAILURE TO OPERATE WITH NO CURRENT DRAW INDICATES: (a) Open field circuit. Inspect internal connections and trace circuit with a test lamp. (b) Open armature coils. Inspect commutator coils. Inspect commutator for badly burned bars.

Running free speed, an open armature will show excessive arcing at commutator bar which is open. (c) Broken or weakened brush springs, worn brushes, high mica on commutator or other causes which will prevent contact between brushes and commutator.

LOW NO-LOAD SPEED, WITH LOW TORQUE AND LOW CURRENT DRAW IS OFTEN DUE TO: (a) An open field winding. Raise and insulate ungrounded brushes from commutator and check fields with test lamp. (b) High internal resistance due to poor connections, defective leads, dirty commutator, broken or weak brush springs, worn brushes or high mica on commutator.

HIGH FREE SPEED WITH LOW TORQUE AND HIGH CURRENT DRAW INDICATES: (a) Shorted fields. There is no way to detect shorted fields, since field resistance is normally very low. If shorted fields are suspected, renew fields and check for improved performance.

83. R&R AND OVERHAUL. To remove the starter motor, first disconnect the battery ground cable. Identify the location of all wires, then disconnect wires from starter relay. Remove two cap screws from end of starter motor and remove the motor.

To disassemble starter, drive pin (12—Fig. FO81) from end of armature shaft and remove starter drive (16 through 23). Loosen clamp screw and remove cover band (6). Separate end plates (2 and 8) and armature (13) from field housing (7).

Inspect brushes and commutator. Brushes should form good contact with commutator, and commutator must be reasonably clean and smooth. If it is not, it should be cleaned or turned down in a lathe. If there are burned bars on commutator, it may indicate open circuited armature coils which will prevent proper cranking. Check for worn, dirty or damaged bearings and armature shaft which could cause starter motor to drag. Renew parts as required.

To install starter motor, position starter on engine and secure with two cap screws. Connect wires to relay switch. Connect battery lead.

IGNITION SYSTEM

84. TROUBLESHOOTING. Most problems involving failure to start, poor or rough running, all stem from trouble in the ignition system. Many novice troubleshooters assume that these symptoms point to the fuel system instead of the ignition system. Note the following symptoms:

    a. Engine misses.

    b. Stumbles on acceleration (misfiring).

    c. Loss of power at high speed (misfiring).

    d. Hard starting (if at all).

    e. Rough idle.

These symptoms may be caused by one of the following:

a. Spark plug(s).
b. Secondary wires (thick wires running from distributor cap to spark plugs and ignition coil).
c. Distributor cap and rotor.
d. Ignition coil.

Ignition system troubles may be roughly divided between those affecting only one cylinder and those affecting all cylinders. If the problem affects only one cylinder, it can only be in the spark plug, secondary wiring or that part of the distributor associated with that cylinder. If the problem affects all cylinders (weak or no spark), then the trouble is in the ignition coil, rotor, distributor or associated wiring.

Perform a spark intensity test as follows: Connect a remote start button to the engine according to manufacturer's instructions. Disconnect one spark plug wire from the plug. Insert a metal adapter in the plug boot and hold it about $3/16$ inch (4.8 mm) from a clean engine ground with insulated pliers. Crank the engine briefly and note the spark intensity. Repeat the procedure to check each spark plug.

If all plug wires deliver a strong bright spark, the secondary circuit is good, but one or more plugs could be weak. Inspect the spark plug condition.

If the spark is weak at some wires and good at others, check the secondary wiring and connections. If the secondary wiring checks out good, check for voltage drop between the battery and coil BAT terminal with the switch in ON position. Normal voltage drop is 0.3 volt. If the voltage drop is normal, the problem is either in the coil or the circuit between the coil and distributor ground.

If the spark is weak at all wires, proceed as follows: Disconnect the coil-to-distributor secondary lead from distributor cap. Using insulated pliers, hold end of wire about $3/16$ inch (4.8 mm) from the engine block. Crank the engine briefly and note the spark intensity. A good spark indicates possible trouble in the secondary circuit. If there is a weak spark or no spark, look for the problem in the coil, primary circuit or the coil-to-distributor secondary lead.

BATTERY IGNITION

Models 9N-2N-Early 8N

85. On 9N, 2N and 8N tractors prior to serial number 263844, the Ford 9N 12100 battery ignition distributor is mounted on the front face of the timing gear cover directly in line with the camshaft. In these installations, the front end of the camshaft is slotted to receive a mating tang on the drive end of the distributor shaft.

The slot in the end of the camshaft is offset so the distributor can only be installed in one position. Thus, the static timing will automatically be correct regardless of crankshaft position when distributor is installed on the engine. Note that the number one cylinder is located at front of engine (closest to radiator).

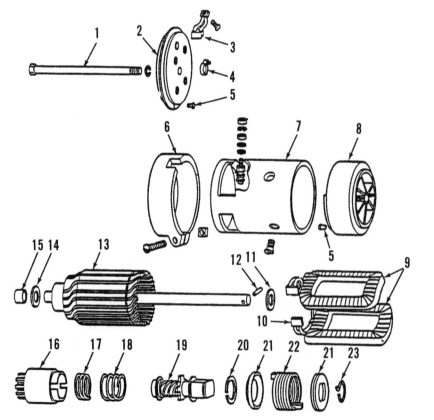

Fig. FO81—Exploded view of starter motor typical of all models.

1. Through-bolt	
2. Brush end plate	13. Armature
3. Brush	14. Thrust washer
4. Brush spring	15. Bearing
5. Locating dowels	16. Starter drive pinion
6. Cover band	17. Meshing spring
7. Field frame	18. Anti-drift spring
8. Rear plate	19. Drive shaft
9. Field coils	20. Retaining ring
10. Brushes	21. Spring anchor plates
11. Thrust washer	22. Starter drive spring
12. Pin	23. Retaining ring

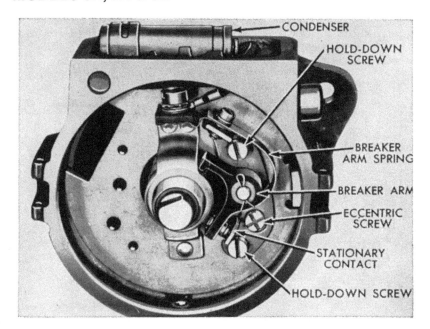

Fig. FO82—View of distributor breaker contacts of Ford 9N12100 distributor used on Models 9N, 2N and early 8N (prior to serial number 263844).

Spark timing at speeds above idling is controlled automatically by centrifugal weights built into the distributor. Full automatic spark advance is 24 crankshaft degrees, 12 distributor degrees at 2000 engine rpm. Recommended contact point gap is 0.015 inch (0.38 mm) or a dwell of 35-38°.

86. BREAKER CONTACTS REPLACEMENT AND ADJUSTMENT. The distributor must be removed to service the breaker points. Disconnect primary wire from the ignition coil and disconnect the spark plug cables from the distributor cap. Unsnap the coil retainer bail and distributor cap clips, then remove coil and cap. Remove the two cap screws attaching the distributor housing to the engine and withdraw the distributor unit.

Loosen the condenser lead and mounting screw and remove the condenser. See Fig. FO82. Remove the hold-down screws and remove the breaker contacts. Install new contact points and secure lightly with the hold-down screws. Make certain that the contact points are parallel when closed. Breaker points that are not properly aligned cause excess heating of the points which results in rapid burning and wear. Rotate the distributor cam until the rubbing block of the breaker arm is positioned at a high point of the cam. Turn the eccentric screw (Fig. FO82) to obtain an air gap of 0.015 inch (0.38 mm) between the contact points. Use a feeler gauge to check the air gap.

CAUTION: Be sure to remove any oil film from the feeler gauge before inserting it between the points. Oil on the contact points will cause rapid burning of the points.

Tighten the breaker contact hold-down screws and recheck the air gap. Install the condenser. Apply a light coat of lubricant to the breaker point cam. Set the distributor basic timing as outlined in paragraph 87. Install distributor on the engine, aligning offset tang and slot. Assemble coil and cap and reconnect wiring.

87. TIMING TO ENGINE. With distributor removed from the engine, adjust the breaker contacts to 0.015 inch (0.38 mm) gap by loosening the two hold-down screws (Fig. FO82) and turning the eccentric screw. Place a straight edge against the tang on the side of the distributor shaft as shown in Fig. FO83, then rotate the shaft until the straight edge is ¼ inch (6.35 mm) from the near edge of the small mounting hole as shown. With shaft in this position, the contacts should be just starting to open. Move the adjustment plate on the left side of the distributor down to advance the timing, up to retard the timing. If proper timing is not obtained on the first attempt, turn shaft backward ½ turn to eliminate backlash, then repeat the check.

NOTE: The engines used in 9N, 2N and early 8N tractors were not provided with timing marks.

Recommended static timing is top dead center which will be automatically obtained regardless of crankshaft position when the distributor is installed in the engine, providing that the basic timing of the distributor was correctly set at ¼ inch (6.35 mm) as outlined above. Recommended running timing fully advanced is 24 flywheel degrees before top dead center when engine is running at 2000 rpm or higher. Any change in the timing to accommodate different fuels is accomplished by changing the basic timing, that is by moving the adjustment plate on the left side of the distributor housing.

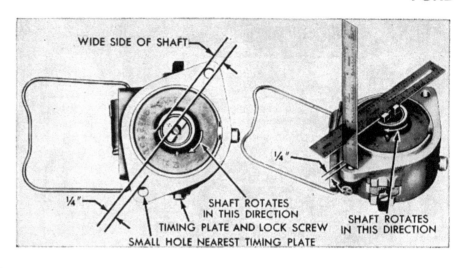

Fig. FO83—Distributor basic timing adjustment on Models 9N, 2N and early 8N (prior to serial number 263844).

Fig. FO84—Angle mount distributor is used on late Model 8N (after serial number 263843).

1. Primary wire
2. Oiler
3. Retaining plate
4. Retaining clips

Engine RPM	0 to 400	1200	2000
Corresponding Spark Advance (Crankshaft Degrees BTCD)	4°	9°-11°	16°-18°

Fig. FO85—Spark advance chart for 8N models with 8N12127 distributor.

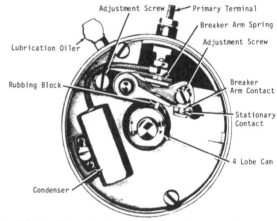

Fig. FO86—View of distributor breaker contacts of Ford 8N12127 distributor used on late Model 8N (after serial number 263843).

BATTERY IGNITION

Model 8N With Angle Drive Distributor

88. Late Model 8N tractors, after serial number 263843, are equipped with an angle-mounted Ford distributor 8N12127A or 8N12127B. See Fig. FO84. The unit is driven by a replaceable gear on the front end of the camshaft and is mounted on the top of the timing gear cover.

Spark timing at speeds above 400 rpm is controlled automatically by centrifugal weights incorporated in the distributor. The centrifugal weights advance the breaker cam as the engine speed increases to allow the spark to occur earlier at the spark plugs. The ignition timing at any given engine speed can be determined on the engine by the use of a timing light. Refer to the spark advance table shown in Fig. FO85.

Recommended breaker contact gap is 0.024-0.026 inch (0.61-0.66 mm). Breaker arm spring tension should be 17-20 ounces (4.7-5.5 N). Firing order is 1-2-4-3.

Apply a few drops of SAE 30 engine oil in distributor oiler (2—Fig. FO84) after every 10 hours of operation.

89. BREAKER CONTACTS REPLACEMENT AND ADJUSTMENT. The distributor does not need to be removed to service the breaker contact points. To renew the breaker points, remove the distributor cap and rotor. Remove the breaker arm spring retaining screw (Fig. FO86) and the two screws that attach the contact points to the breaker plate. Lift out the contact points and spring. Install new contact points in normal operating position and secure lightly with the retaining screws. Install the condenser and the breaker arm spring screw. Make certain that the contact points are parallel when closed. Points that are not properly aligned cause excess heating of the points which results in rapid burning and wear. Rotate the distributor cam until the rubbing block of the breaker arm is positioned at a high point of the cam.

Fig. FO87—On 8N models with angle mount distributor, the flywheel is stamped with timing marks from 0 to 20 degrees. Timing marks may be viewed through timing hole in right side of bell housing (1).

Move the stationary contact point as necessary to obtain an air gap of 0.0024-0.0026 inch (0.61-0.66 mm) between the contact points. It is recommended that a feeler gauge be used to check the air gap.

> CAUTION: Be sure to remove any oil film from the feeler gauge before inserting between the points. Oil on the contact points will cause rapid burning of the points.

Tighten the breaker contact retaining screws and recheck the air gap. Apply a light coat of lubricant to the breaker point cam.

> NOTE: Ignition timing should be checked each time the ignition points are adjusted or renewed.

90. TIMING TO ENGINE. The flywheel of production engines equipped with angle drive distributors is stamped with an "O" mark indicating top center position and with additional stamped lines up to 20° BTDC in one degree increments. The degree lines index with a pointer at the flywheel timing port located on the right side of the engine as shown in Fig. FO87. The "O" mark and degree lines appear on the flywheel twice, 180° apart, so running timing can be checked from any spark plug wire.

To set the basic ignition timing, loosen timing hole cover retaining screw and move the cover to expose timing port (Fig. FO87). Remove the spark plug from No. 1 cylinder. Note that the number one cylinder is located at front of engine. Remove the distributor cap. Rotate the crankshaft until compression pressure is felt at the No. 1 spark plug hole. Continue turning the crankshaft, using a screwdriver in the flywheel ring gear teeth, until the 4° flywheel timing mark is aligned with the pointer in the timing hole. In this position, the ignition contact points should just begin to open. If the ignition points must be moved to obtain the correct separation point,

loosen the clamp screw (3—Fig. FO84) on the distributor retaining plate and rotate the distributor body as necessary.

Using a timing light, recheck the basic timing and the centrifugal advance range as follows: Start the engine and run at 400 rpm. Direct the timing light on the flywheel at the timing port and note the timing marks. Timing should be 4° before top dead center. To advance the timing, turn the distributor housing clockwise; to retard the timing, turn the distributor housing counterclockwise. Increase engine speed to 1200 rpm and 2000 rpm while observing the timing marks. Refer to table in Fig. FO85 for specified spark advance curve.

91. R&R DISTRIBUTOR. To remove the distributor, first remove the timing hole cover from right hand side of the bell housing (Fig. FO87). Remove the No. 1 spark plug. Note that the number one cylinder is located at front of engine (closest to radiator). While holding thumb over No. 1 spark plug hole, turn the crankshaft until compression pressure is felt at the spark plug hole (No. 1 piston is on the compression stroke). Continue turning crankshaft until the 4° timing mark on flywheel is aligned with the timing pointer on the bell housing. Scribe a mark on the distributor body in line with the No. 1 spark plug cable in the distributor cap. Remove the distributor cap. Disconnect the primary wire (1—Fig. FO84) from the distributor. Remove the cap screw (3) from the retainer plate and lift the distributor from the engine front cover.

See Fig. FO88 for an exploded view of distributor. Make certain that cam weights (21) move freely. Check distributor shaft (20) for excessive end play or side-to-side play. If a new distributor shaft is being installed, hole must be drilled in shaft for the collar retaining rivet (9) after shaft, thrust washer (2), bushing (3) and collar (8) are assembled in distributor housing. Position collar (8) on shaft and drill hole for rivet so that shaft will have end play of 0.003-0.009 in. (0.08-0.23 mm).

To reinstall the distributor, turn the distributor shaft until the rotor is in line with the scribe mark on the distributor body. Make sure that the No. 1 piston is on the compression stroke and the 4° timing mark on flywheel is aligned with the mark on the bell housing. Insert the distributor shaft into the front cover. Turn the distributor body until the points just begin to open, then tighten the retaining plate cap screw. Install the distributor cap, making sure that the rotor points to the No. 1 spark plug cable.

Check the ignition timing with a timing light and complete the timing as outlined in paragraph 90.

MAGNETO IGNITION SYSTEM

Model 2N

92. The magneto ignition system was a wartime contingency aimed at providing some tractor production starting in 1942, and very few tractors were actually so equipped. A Fairbanks-Morse FM-J4B73 magneto unit is used (Fig. FO89), and is driven by the camshaft through a magneto drive adapter as shown in Fig. FO90.

93. MAGNETO TIMING AND ADJUSTMENT.

The magneto mounting flange and mating adapter

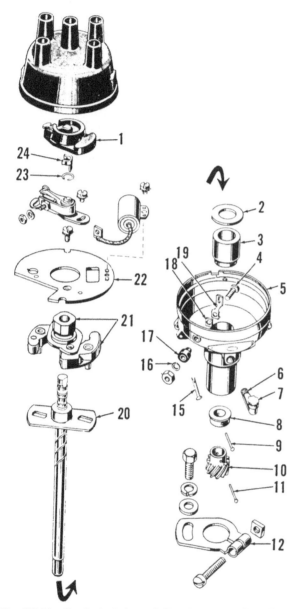

Fig. FO88—Exploded view of distributor used on 8N tractors after serial number 263843.

1. Rotor	12. Retainer plate
2. Thrust washer	15. Rivet
3. Bushing	16. Washer
4. Primary screw	17. Bushing
5. Body	18. Insulator
6. Wick	19. Primary terminal
7. Oiler	20. Distributor shaft
8. Collar	21. Cam & weight assy.
9. Pin	22. Breaker plate
10. Drive gear	23. Cam retainer
11. Pin	24. Spring

flange are provided with slotted holes (2—Fig. FO89) to permit timing adjustment. The magneto uses an impulse coupling with a 15° lag angle, which automatically advances the spark timing that amount when the engine starts to run. For timing purposes, the impulse coupling should trip when the No. 1 piston is at top dead center of compression stroke.

Check and align the contact points and adjust gap to 0.020 inch (0.5 mm) when contacts are wide open. To adjust, loosen lock screw (20—Fig. FO91) located at upper end of stationary contact plate, then turn eccentric adjusting screw (18) to obtain specified contact gap setting. Tighten lock screw and recheck gap spacing. Excessive rubbing block wear will cause internal magneto timing to vary. Rubbing block wear is considered excessive when contour of block matches that of the cam.

Running test spark should jump ¼ inch (6 mm) gap. Condenser capacity is 20 MFD plus or minus 10 percent.

94. MAGNETO TEST AND OVERHAUL.

Before removing magneto from tractor as a suspected source of trouble, the unit should be checked to determine whether it is faulty. To determine magneto general condition when no spark is apparent at spark plugs, remove spark plug cables at the distributor block and attach a piece of stiff wire at each terminal. Bend attached wires until the free ends are approximately ¼ inch (6 mm) from a grounded surface to form individual, external spark gaps. Then move magneto switch to ON position and turn engine over slowly, observing whether a spark occurs at one of the spark gaps each time the impulse coupling releases.

If sparks of equal intensity are produced at all gaps, the magneto is in good condition and remainder of high tension system must be checked for defects. If no sparks are produced at spark gaps, inspect the

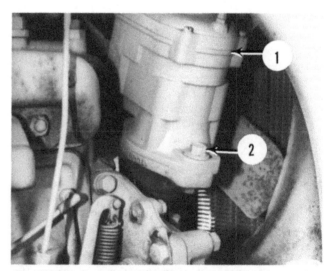

Fig. FO89—A magneto ignition system (1) was used on some Model 2N tractors. Slotted mounting hole (2) permits timing adjustment. Refer to text.

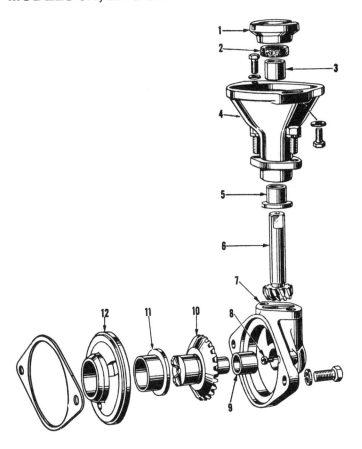

Fig. FO90—Magneto drive adapter was a wartime replacement option on some 2N models.

1. Drive coupling
2. Oil wick
3. Bushing
4. Upper housing
5. Bushing
6. Driven gear
7. Lower housing
8. Plate screw
9. Bushing
10. Drive gear
11. Bushing
12. Adapter plate

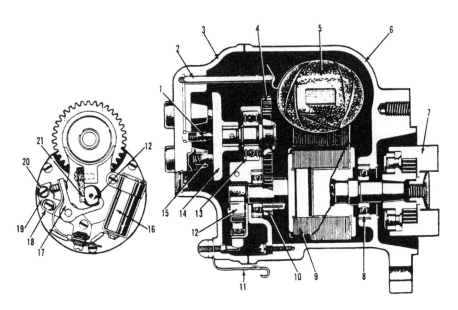

Fig. FO91—Fairbanks-Morse Type FM magneto side cross section and breaker assembly.

1. Center brush
2. High tension lead
3. End cap
4. Distributor gear
5. Coil
6. Housing
7. Impulse coupling
8. Rotor thrust bearing
9. Magnet rotor
10. Rotor gear
11. Grounding switch
12. Breaker cam
13. Ball bearing
14. Distributor rotor
15. Terminal brushes
16. Condenser
17. Breaker arm
18. Eccentric screw
19. Contact plate
20. Lock screw
21. Oiler wick

breaker contacts for condition and spacing, also check magneto switch and circuit for high resistance or grounds. In an emergency, the coil or condenser may be checked by the comparison method, provided the parts can be readily removed. Use parts that are know to be in good condition in place of parts suspected as defective. If no discrepancies are found by the outlined checks and no sparks are produced, the magneto must be removed from the tractor for shop repair.

Disassembly sequence for magneto is basically as follows: Impulse coupling, distributor compartment cover, end cap and rotor, breaker contacts and condenser.

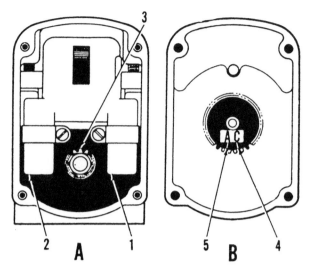

Fig. FO92—Determining rotation of Fairbanks-Morse typical type FM-J magneto.

   A. To distinguish direction of rotation
   1. On clockwise units condenser is mounted in this position
   2. On counterclockwise units condenser is mounted in this position
   3. Marked tooth on magnet rotor
   B. Meshing distributor gear and magnet rotor pinion
   4. "C" for clockwise rotation
   5. "A" for counterclockwise rotation

NOTE: Do not attempt to remove the magnetic rotor from the frame unless specific instructions are available.

If necessary to replace rotor thrust bearing (8—Fig. FO91), the insulating strip and washers should always be renewed. The bearing is electrically insulated from the frame and should be tested for leakage after installation, using test lamp and test points. Repack ball bearings, using a high consistency and a very high melting point grease. Oilite bushing grooves are refilled with a light, low melting point grease. If cam felt wick (21) is hardened or dried, renew wick. Oilite bushings should be installed, using close fitting pilots to maintain proper dimensions and to avoid reaming. Observe bushing installation before removing. Install new bushing with ends set flush or protruding as required.

Magneto rotation may be determined by observing location of condenser (A—Fig. FO92) as follows: Clockwise (CW) rotation, condenser is mounted to the right of the breaker assembly, and to left of breaker assembly for counterclockwise (CCW) rotation.

Examine parts for damage or excessive wear. Check coil for continuity, short circuits, high resistance, output and high voltage leaks. Test condenser for breakdown, leakage, capacity and series resistance. Inspect breaker contacts for being pitted or burnt and check breaker arm rubbing block for excessive wear. Resurface breaker contacts, using a fine abrasive oil stone or renew them if badly burnt or worn. Examine distributor parts for evidence of high voltage leakage and give electrical test for leakage or insulation breakdown.

Air ventilating passages in distributor end cap should be kept clean. New gaskets should be used when reassembling. Gasket joints should be sealed, using gasket sealing compound, as moisture condensation inside the magneto is usually caused by improper sealing.

Distributor discs and end caps should not be cleaned with sand paper or emery cloth. Remove carbon deposits using clean cloth, moistened in cleaning solvent. Mesh timing marks on rotor pinion and gear, when assembling as follows: For counterclockwise (CCW) rotation single beveled tooth of pinion meshes with "A" of distributor gear (Fig. FO92).

Clean impulse coupling in solvent and use engine oil to lubricate.

95. DRIVE ADAPTER OVERHAUL. The adapter is mounted on the front of the timing gear cover in the same manner as the battery ignition distributor. Remove the adapter and proceed to disassemble, using care to mark the gears and couplings to assure proper assembly. Refer to Fig. FO90. Renew excessively worn parts, reassemble, install and retime ignition as previously described.

CLUTCH

TROUBLESHOOTING

All Models

96. Several clutch problems may be experienced. Usually the trouble is quite obvious and will fall into one of the following categories:
1. Slipping, chattering or grabbing when engaging.
2. Spinning or dragging when disengaged.
3. Clutch noises, clutch pedal pulsations and rapid clutch disc facing wear.

Clutch Slips While Engaged. Clutch linkage is improperly adjusted, the pressure springs are weak or broken, the friction disc facings are worn or the disc is contaminated with grease or oil.

This problem is most noticeable when pulling a heavy load in a high gear. To check slippage, park the tractor on a level surface with the brakes applied. Shift to second gear and release the clutch as if driving off. If the clutch is good, the engine will slow and stall. If the clutch slips, continued engine speed will give it away.

Clutch Chatters or Grabs When Engaging. Clutch linkage is improperly adjusted, the friction disc facings are contaminated with grease or oil, or clutch components are worn and/or damaged.

Clutch Spins or Drags When Disengaged. The clutch friction disc normally spins briefly after disengagement and takes a moment to come to rest. The sound should not be confused with drag.

Drag is caused by the friction disc not being fully released from the flywheel or pressure plate as the clutch pedal is depressed. It usually causes difficult shifting and gear clash. This problem can be caused by improperly adjusted linkage or defective/worn clutch components.

Clutch Noises. Clutch noise is generally most noticeable when the engine is idling. Note whether the noise is heard when the clutch is engaged or disengaged. Clutch noises when engaged could be caused by a loose friction disc hub, loose disc springs and misalignment or looseness of the transmission input shaft. When disengaged, noises can be caused by a worn release bearing, defective pilot bearing or a misaligned release lever.

Clutch Pedal Pulsates. This problem is generally noticed when slight pressure is applied to the clutch pedal with the engine running. As pedal pressure is increased, the pulsation ceases. Possible causes include a bent crankshaft flange, distortion or shifting of the clutch housing, a misaligned release lever, warped friction disc or a damaged pressure plate.

Rapid Friction Disc Facing Wear. This problem is caused by any condition that permits slippage between the facings and the flywheel or pressure plate. Probable causes are "riding" the clutch, slow releasing of the clutch after disengagement, weak or broken pressure springs, improperly adjusted pedal linkage and a warped clutch disc or pressure plate.

ADJUSTMENT

Models 9N-2N

97. Free travel of the clutch pedal is the only adjustment necessary for proper operation of the clutch. Free travel is the distance the clutch pedal can be pushed down before resistance is met (the moment the clutch release bearing contacts the clutch release levers). Pedal free travel should be $3/16$ inch (4.8 mm), measured between top of pedal and pedal stop (Fig. FO93). Pedal should have additional travel of $19/16$ inches (40 mm) before pedal contacts brake pawl arm.

To adjust the clutch pedal free travel, disconnect the clutch rod clevis (Fig. FO93) and turn the clevis to lengthen the rod (increase free travel) or shorten the rod (decrease free travel).

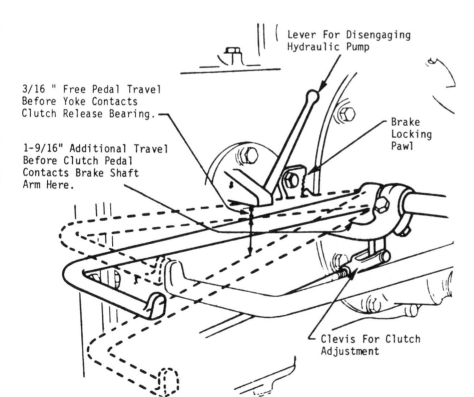

Lever For Disengaging Hydraulic Pump

3/16 " Free Pedal Travel Before Yoke Contacts Clutch Release Bearing.

1-9/16" Additional Travel Before Clutch Pedal Contacts Brake Shaft Arm Here.

Brake Locking Pawl

Clevis For Clutch Adjustment

Fig. FO93—Clutch pedal free travel adjustment on Models 9N and 2N.

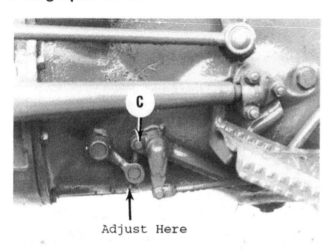

Fig. FO94—To adjust clutch pedal free travel on 8N tractors, remove pin (C) and rotate the eye-bolt.

Model 8N

98. Free travel of the clutch pedal is the only adjustment necessary for proper operation of the clutch. Free travel is the distance the clutch pedal can be pushed down before resistance is met (clutch release bearing contacts the clutch release levers). Pedal free travel should be ¾ inch (19 mm) measured at the pedal foot pad.

To adjust the clutch pedal free travel, remove the clevis pin from the clutch lever eye bolt as shown in Fig. FO94 and turn the eye bolt until the adjustment is correct.

TRACTOR FRONT SPLIT

Models 9N-2N

99. Shut off fuel, then disconnect fuel line from the carburetor. Remove four cap screws that secure hood to instrument panel, and remove two cap screws securing hood to front axle support. Lift the hood and fuel tank from the tractor. Block up tractor under transmission case. Support rear end of engine on a wheeled jack. Disconnect both radius rods and drag links at rear ends. Tag wires for identification, then disconnect battery, starter, generator and ignition wires or cables. Disconnect throttle and choke controls, oil pressure gauge line at cylinder block, air inlet tube at carburetor and exhaust pipe at manifold. Remove bolts holding transmission case to engine and separate engine and transmission as shown in Fig. FO95.

Reconnecting the tractor is the reverse of splitting procedure, plus the following steps. If the clutch assembly was removed from the flywheel, use a short dummy clutch shaft or other suitable aligning tool to align the clutch hub with the flywheel pilot bearing. See Fig. FO96. The engine and transmission should slide together easily when properly aligned. Note that it may be necessary to rotate the flywheel to align splines of clutch hub with splines of transmission input shaft. Do not tighten the transmission-to-engine mounting bolts until engine and transmission housings are completely together. Tighten transmission mounting bolts to 53-60 ft.-lbs. (72-81 N•m) torque.

Fig. FO95—Front assembly separated from transmission on Models 9N-2N-8N.

Fig. FO96—Use wooden wedges and an alignment tool to facilitate removal and installation of clutch.

Model 8N

100. The 8N tractor can be split between the engine and transmission without removing the hood and fuel tank, if desired, as follows: Disconnect the battery cables and remove the battery. Disconnect the starter button wire. It is not necessary to disconnect the starter, generator or ignition wiring as the instrument panel and steering gear remain with the engine in this splitting procedure. However, it is recommended that the throttle and choke linkage, air cleaner hose, fuel line and oil pressure gauge line be disconnected to avoid binding when the steering gear is raised to clear the transmission housing. Disconnect the exhaust pipe at the muffler. Disconnect the front axle radius rods at the rear ends. Remove cap screws attaching steering gear to transmission housing. Place wood wedges between the front axle and front support to prevent tipping. Block up the front of the tractor under the oil pan. Support the rear of the tractor with a rolling floor jack under the transmission housing. Remove the bolts attaching the transmission housing to the engine. Attach an overhead hoist with a rope or chain passing through the steering wheel, then raise the steering gear until clear of the transmission housing. If an overhead hoist is not available, the hood and steering gear can be raised manually by lifting on ends of front axle radius rods so that steering gear is clear of transmission housing. Place a block of wood between bottom of battery box and rear of engine to support steering gear in raised position. Roll the rear half of the tractor away from the engine.

Reconnecting the tractor is the reverse of splitting procedure, plus the following steps. If the clutch assembly was removed from the flywheel, use a short dummy clutch shaft or other suitable aligning tool to align the clutch hub with the flywheel pilot bearing. See Fig. FO96. The engine and transmission should slide together easily when properly aligned. Note that it may be necessary to rotate the flywheel to align splines of clutch hub with slines of transmission input shaft. Do not tighten engine-to-transmission mounting bolts until engine and transmission housings are completely together. Tighten transmission mounting bolts to 53-60 ft.-lbs. (72-81 N•m) torque.

OVERHAUL

All Models

101. To remove the clutch, first split the tractor between the engine and transmission (paragraph 99 or 100). With clutch assembly accessible, mark clutch cover and flywheel to assure correct balance when reinstalling. Force clutch release levers inward and insert wooden wedges between levers and cover as shown in Fig. FO96. Unscrew cap screws holding clutch to flywheel and remove the clutch cover assembly and lined disc from the flywheel.

The clutch pressure plate and cover unit is serviced as an assembly only and exchange units are supplied at most places serving the market. Renew the assembly if pressure plate is scored, warped, cracked or excessively worn, or if the cover release fingers or springs are damaged. The clutch disc is also serviced as an exchange unit and the cover and clutch disc are normally renewed as a service unit.

NOTE: Both the clutch cover assembly and lined friction disc are available as regular or heavy duty units. Like units (regular or heavy duty) should be serviced as a matched pair, that is, a heavy duty friction disc should be used with a heavy duty cover and a regular duty friction disc should be used with a regular duty cover.

When servicing the clutch, inspect clutch pilot bearing in the flywheel and renew or service as necessary. Inspect the clutch throwout bearing and renew if its condition is questionable. The throwout bearing is a prelubricated assembly and cannot be serviced. Do not wash bearing assembly in solvent or attempt to lubricate with oil.

Install clutch disc with long side of disc hub facing away from flywheel. Use a short dummy clutch shaft or other suitable aligning tool to align the clutch hub with the flywheel pilot bearing. See Fig. FO96.

NOTE: If a suitable clutch alignment tool is not available, the main clutch shaft and retainer can be removed from the transmission by first removing the clutch release bearing and springs, then remov-ing the four cap screws which secure the retainer. Keep all parts together and reinstall as soon as clutch is mounted on the flywheel. There should be only one gasket on the retainer.

Position clutch pressure plate on flywheel, aligning match marks made prior to removal, and tighten mounting cap screws evenly. Remove wooden wedges from between release levers and cover.

SHERMAN TRANSMISSION

102. Some tractors have been equipped with a Sherman Combination Transmission as a field installation. This three-range (over, under and direct drive) transmission is installed between the engine clutch and the main transmission.

When the Ford tractor was introduced, it was designed for drawbar work at full throttle (2000 engine rpm; 800 pto rpm), and pto operation at ¾ throttle (1500 engine rpm). This theory was not universally accepted by tractor owners who wanted to operate at full throttle even when using the pto. Sherman Products Company introduced the Sherman step-up transmission which permitted proper use of the pto at full engine speed, and also provided three additional working speeds. The success of this accessory was quickly followed by the very popular Sherman Combination Transmission which provided three pto speeds as well as nine forward and three reverse ground speeds.

A similar unit was marketed by the Hupp Motor Company, with sales mostly directed toward the newly franchised Ferguson Tractor Dealers, but the units were interchangeable and some have been installed in Ford tractors.

NOTE: Service parts for these transmissions are not available from Ford.

REMOVE AND REINSTALL

Models So Equipped

103. Separate (split) the tractor between the engine and transmission as outlined in paragraph 99 or 100. Disconnect the clutch release bearing spring and remove the bearing. Remove the left brake pedal and the brake cross shaft. Remove the socket head screw from the Sherman transmission shift lever and remove the lever, then unbolt and remove the Sherman transmission from the mounting flange on the front face of the main transmission housing. The Sherman transmission mounting flange can be removed at this time by removing the four socket head screws and pulling the flange and shims from the tractor transmission housing as shown in Fig. FO97.

Installation is the reverse of removal procedure plus the following steps. Check drive shaft bearing preload and adjust if necessary. Fill unit with Ford M2C 134-C fluid, then install the clutch shaft assembly. Reconnect the engine and transmission as outlined in paragraph 99 or 100.

104. BEARING PRELOAD ADJUSTMENT. Check the bearing preload of the Sherman transmission as follows: Install approximately 0.050 inch (1.27 mm) of shims behind the mounting flange as shown in Fig. FO97. Be sure that the flat side of the mounting flange is at the top. Install the Sherman transmission with a new mounting flange gasket. Note that the transmission drive (output) shaft should have some end play at this time. Remove shims from behind the mounting flange until the drive shaft has zero end play, then remove one additional 0.007 inch (0.18 mm) shim to obtain the desired prelaod of the drive shaft bearing.

OVERHAUL

Models So Equipped

105. SHIFTER RAILS AND FORKS. To remove the shifter rails and forks, the Sherman transmission

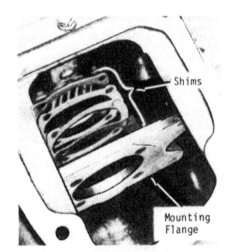

Fig. FO97—View showing Sherman transmission mounting flange removed from front wall of tractor transmission. Note shims between the flange and transmission which control preload of Sherman transmission output shaft and bearings.

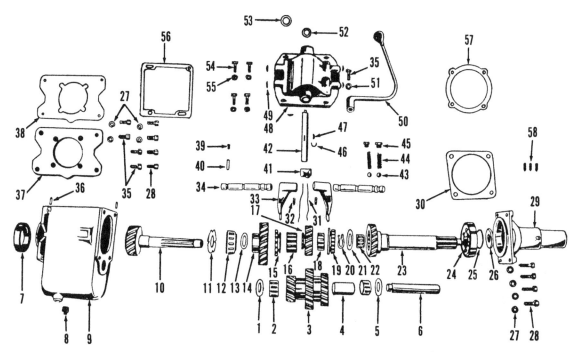

Fig. FO98—Exploded view of the Sherman transmission showing the component parts.

1. Thrust washer
2. Bearing
3. Cluster gear
4. Bearing spacer
5. Thrust washer
6. Cluster shaft
7. Bearing cup
8. Drain plug
9. Gear case
10. Drive (output) shaft
11. Oil slinger
12. Bearing

13. Gear washer
14. Step-down gear
15. Rear shift collar
16. Spline sleeve, long
17. Step-up gear
18. Spline sleeve, short
19. Front shift collar
20. Snap ring
21. Thrust washer
22. Needle bearing
23. Clutch (input) shaft
24. Ball bearing

25. Snap ring
26. Oil seal
29. Support assy.
32. Drilled set screw
33. Shifter finger
34. Shifter rod
36. Dowel
37. Mounting flange
38. Flange gasket
40. Interlock
41. Selector fork

42. Selector shaft
43. Detent balls
44. Detent spring
45. Detent cap
46. Snap ring
47. Woodruff key
48. Shifter cap
49. Welch plugs
50. Shift lever
52. Seal
53. Rubber grommet
57. Shim

must be removed from the tractor as outlined in paragraph 103.

Refer to Fig. FO98 and remove the retaining cap screws. Lift the shifter cap assembly (48) from the gear case. Remove the two detent caps (45), detent springs and detent balls from shifter fingers (forks) (33). Remove the Welch plugs (49) from the shifter rod bores at the rear of the shifter cap, then drive the shifter rods (34) out the front of shifter cap. Remove the set screw from the interlock bore and remove the interlock (40). Remove the snap ring from the groove on the selector shaft (42) and slide the selector fork (41) and snap ring (46) toward the end of the selector shaft, which is opposite from the shoulder stop. Remove the Woodruff key (47) from the selector shaft then withdraw the selector shaft from the shifter cap (48). Remove selector shaft seal (52) from the bore in shifter cap.

Wash all parts with a suitable solvent, then inspect as follows: Position the shifter rods (34—Fig. FO98) and selector shaft (42) in their respective bores and check for freedom of movement. If binding occurs due to rods or shaft being bent, renew the bent part. If binding occurs due to rods or shaft being scored, it may be possible to recondition the scored part using

Crocus cloth. Inspect the shifter fingers (33) and selector fork (41) and renew them if they show signs of contact at points other than the contact pads. Inspect the interlock (40) for flat spots or signs of scoring, and renew if necessary.

When reassembling, use new seal (52—Fig. FO98) and Welch plugs and proceed as follows: Place the selector shaft oil seal (52) in its bore and drift into position using a driver of the proper size. Start the selector shaft in its bore and place the selector fork (41) and snap ring (48) on shaft. Install Woodruff key in its slot, then slide selector shaft (42) into place and position snap ring in its groove. Start a shifter rod (34) in the bore farthest (opposite) from interlock plug and be sure the end with the two grooves is toward the detent end of shifter cap. Place a shifter finger (fork) (33) over shifter rod so that the recess in shifter finger will engage selector fork (41).

NOTE: The shifter fingers (forks) and shifter rods are identical and can be interchanged.

Position parts and align the center groove in the shifter rod (34) with the tapped hole in the shifter finger (33), then secure shifter finger with the drilled

Fig. FO99—View showing clutch shaft and support assembly removed from Sherman transmission gear case.

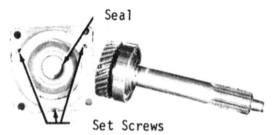

Fig. FO100—Clutch shaft and bearing assembly can be removed from front support after loosening Allen screws in support.

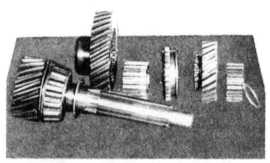

Fig. FO101—View showing drive (output) shaft partially disassembled. Note the position of the long and short spline sleeves.

shaft (23) for chipped teeth, cupping of the pilot bore, scoring or other signs of wear or damage. Check thrust washer (21) for scoring or undue wear. Renew parts as needed.

When reassembling, use a new oil seal (26) in front support. Press bearing (24) on clutch shaft and secure with a snap ring (25). Install the clutch shaft and bearing in the front support and secure with the three socket head screws (58). Fill unit with Ford M2C 134-C oil. Place the thrust washer (21) first, then the roller bearing (22) on the pilot end of drive shaft (10). Using a new gasket, install the front support and clutch shaft assembly after making certain that the oil return hole in front support is on the bottom.

107. DRIVE (OUTPUT) SHAFT. The drive shaft can be removed after the Sherman transmission is removed from the tractor transmission and the shifter assembly and clutch shaft assembly are removed as outlined in paragraphs 106.

Refer to Fig. FO98 and remove the snap ring (20) from drive shaft (10); then remove front shift collar (19), short spline sleeve (18), step-up gear (17), rear shift collar (15), long spline sleeve (16), step-down gear (14) and washer (13) from drive shaft (10). See Fig. FO101.

NOTE: Identify shift collars (15 and 19—Fig. FO98) so they can be reinstalled in their proper positions.

Pull the drive shaft (10), oil slinger (11) and taper bearing (12) from rear of gear case; then, using a suitable press or puller, remove the taper bearing and oil slinger from the drive shaft. Remove bearing cup (7) from gear case if necessary. Clean all parts in a suitable solvent and inspect the taper bearing (12) for rough spots, freedom of movement or other signs of wear or damage. Inspect all drive shaft gears for erratic wear patterns, damaged teeth or splines, cupping, overheating on the gear end faces or scoring on inside diameters and renew as necessary. To install drive shaft, place oil slinger (11) on drive shaft and then place on the taper bearing.

head cap screw (32). Install the interlock and tighten socket head screw (39). Install the other shifter rod (34) and shifter finger (33) in the same manner as the first was installed, then install lock wires (31) in the drilled head cap screws (32) and the holes provided in the shifter fingers (33). Place the detent balls (43) and springs (44) in their bores and install the detent caps (45). Install new Welch plugs. Use a new gasket and install shifter cap assembly to gear case and make sure that shift fingers engage the shift collars. Reinstall Sherman transmission to tractor transmission.

106. CLUTCH (INPUT) SHAFT. With tractor separated (split) as outlined in paragraph 99 or 100, and the unit drained, the clutch shaft (23—Fig. FO98) can be removed by unbolting the support assembly (29) and pulling the support assembly and clutch shaft from the gear case.

Remove the roller bearing (22) and the thrust washer (21) from the pilot end of drive shaft (10). See Fig. FO99. Remove the three socket head screws (58—Fig. FO98) from front support (29) and remove the clutch shaft assembly from support. See Fig. FO100. Remove the snap ring (25—Fig. FO98) and press the ball bearing (24) from clutch shaft. Remove the oil seal (26) from front support using an Owatonna No. 956 bearing puller and slide hammer, or its equivalent.

Clean all parts with a suitable solvent and check the ball bearing (24) for rough spots, flat spots, freedom of movement or other signs of wear. Check clutch

Place shaft and bearing in gear case and install parts as follows: Install gear washer (13) and step-down gear (14) with hub of gear facing rearward. Install long spline sleeve (16), then position the rear shift collar (15) with teeth toward the rear. Install step-up gear (17) with the shift collar engaging teeth rearward. Install the short spline sleeve (18), then position the front shift collar (19) with engaging teeth toward front. Install snap ring (20). Install the shifter cap assembly.

NOTE: The drive shaft bearing preload must be established in the Sherman transmission as outlined previously in paragraph 104.

108. CLUSTER GEAR SHAFT. The cluster gear (3—Fig. FO98) and shaft (6) can be removed after the shift cap, clutch shaft and drive shaft assemblies are removed as outlined previously in paragraphs 105, 106 and 107. Drive the cluster shaft (6) toward rear

of transmission case and withdraw the cluster gear (3) and thrust washers (1 and 5) as the shaft is removed. Remove the roller bearings (2) and the spacer (4) from the inside diameter of the cluster gear.

Check cluster gear for chipped teeth, cupping of end faces, signs of overheating and wear of inside diameter. Check roller bearings for flat spots and freedom of movement or other signs of wear. Check cluster shaft for pitting, scoring or wear of outside diameter. If wear or damage is found on any of the parts, it is recommended that the cluster shaft, gear and bearings be renewed as an assembly.

The procedure for reassembly is the reverse of disassembly; however, keep in mind that the larger of the two end gears of the cluster assembly is toward the front of the gear case. Reassemble the balance of the transmission as outlined previously in paragraphs 105, 106 and 107.

STANDARD TRANSMISSION

REMOVE AND REINSTALL

All Models

109. To remove the transmission as a unit from the tractor, first drain the oil from the transmission and rear axle center housings. Note that a common reservoir is used on 9N, 2N and 8N tractors; however there are three drain plugs which must be removed when draining the fluid.

Split the tractor between the engine and transmission as outlined in paragraph 99 or 100. Unbolt and remove steering gear, instrument panel and battery box (9N and 2N). Remove the step plates from both sides of the transmission housing. Disconnect the left hand brake control rod and the clutch linkage (on 9N and 2N). Remove the muffler. Remove the power take-off control from the left side of the rear axle center housing. Remove the four cap screws attaching the power take-off output shaft retainer housing to rear axle center housing and withdraw the power take-off shaft from the tractor. (Refer to paragraph 139.) Block up the rear axle center housing. Support the transmission housing with a suitable hoist or jack. Remove the bolts attaching the transmission housing to the rear axle center housing, then withdraw the transmission housing from rear housing.

Installation of the transmission is the reverse of removal procedure. Tighten transmission mounting bolts to 53-60 ft.-lbs. (72-81 N•m) torque. Reinstall the power take-off after the transmission housing is reconnected to the rear axle center housing.

OVERHAUL

Models 9N-2N (3 Speed)

110. SHIFT RAILS AND FORKS. It is not necessary to remove the transmission from the tractor in order to remove the shift rails and forks. Rails and forks can be removed from the transmission housing after first separating the rear axle center housing from the transmission housing as outlined in paragraph 126 and removing the steering gear housing and gear shift cover assembly.

Remove the gear shift lock plate (10—Fig. FO102). Pull the shift rails (11) out rearward and lift out the forks (13 and 14). Be careful not to lose the detent balls and springs (12 and 15). Renew worn or damaged parts as necessary.

Reinstall the shift rails and forks in reverse order of their removal.

111. MAINSHAFT. The mainshaft (8—Fig. FO103) is removed from the rear of the transmission housing as follows after the shift rails and forks are removed. Remove the rear bearing retainer (13—Fig. FO103 or FO104) and shims (12). Move the mainshaft toward the rear and remove the pilot bearing cone (6) from the front of the shaft. Pull the mainshaft out the rear of the housing and lift the sliding gears (7 and 9) out through the top opening. Remove the bearing cone (10—Fig. FO105) from the rear of the shaft if necessary.

Clean and inspect the mainshaft and gears and renew worn or damaged parts as needed. Coat all parts with transmission oil during assembly. Assem-

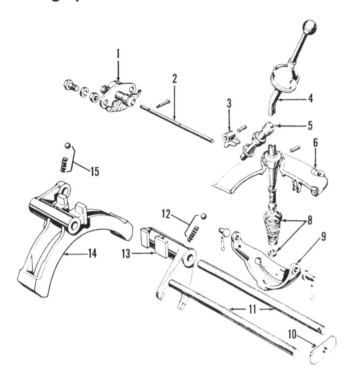

Fig. FO102—Gear shifter mechanism used on Models 9N and 2N. Some changes were made in the design beginning with serial number 12500.

1. Starter switch	9. Shift lever lock
2. Starter push rod	10. Shift rail lock plate
3. Starter rod lever	11. Shifter rails
4. Shifter lever	12. Detent ball & spring
5. Foot button	13. 2nd & reverse shift fork
6. Cover	14. 1st & 3rd shift fork
8. Lever spring & retainer	15. Detent ball & spring

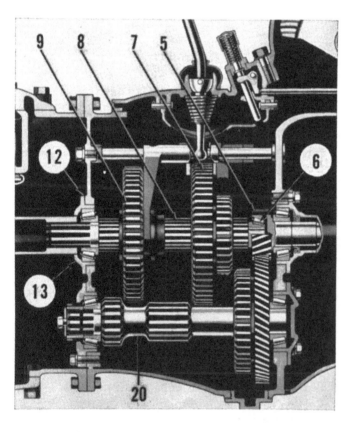

Fig. FO103—Sectional view of transmission used on Models 9N and 2N.

5. Main drive gear	9. 2nd & reverse gear
6. Pilot bearing	12. Shims
7. 1st & 3rd gear	13. Bearing retainer
8. Mainshaft	20. Countershaft

ble the sliding gears (7 and 9—Fig. FO103 or FO105), mainshaft (8) and bearings (6 and 10) into the transmission housing. Adjust mainshaft bearing preload as outlined in paragraph 112.

112. MAINSHAFT BEARING ADJUSTMENT. Install the rear bearing retainer (13—Fig. FO103 or

FO104) with original shims (12) and adjust the bearings as follows: The mainshaft bearings are adjusted by varying the number and thickness of shims (12) under the rear bearing retainer. When bearings are correctly adjusted and the sliding gears in neutral position, 20-25 in.-lbs. (2.3-2.8 N·m) of torque will be required to turn the mainshaft at the rear or output

Fig. FO104—Rear view of 9N and 2N transmission showing shift forks, mainshaft bearing retainer and countershaft bearing retainer removed from transmission housing.

10. Shift rail lock plate
11. Shifter rails
12. Shims
13. Mainshaft bearing retainer
14. 1st & 3rd fork retainer
23. Shims
28. Clutch release lever
29. 2nd & reverse fork
30. Detent ball & spring
31. Pto hub
32. Countershaft bearing support
33. Pto shifter rail

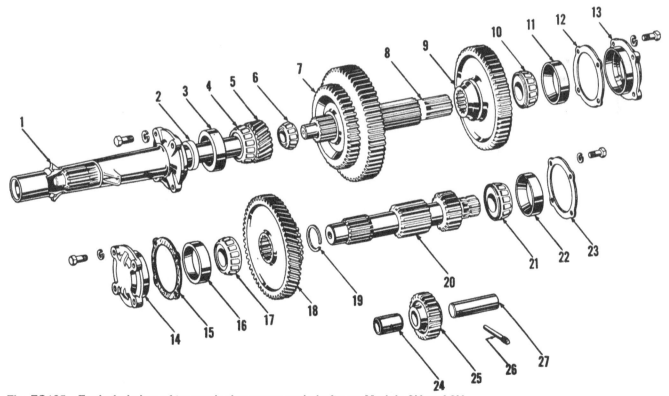

Fig. FO105—Exploded view of transmission gears and shafts on Models 9N and 2N.

1. Bearing retainer	8. Mainshaft	15. Gasket	22. Bearing cup
2. Oil seal	9. 2nd & reverse gear	16. Bearing cup	23. Shims
3. Bearing cup	10. Bearing cone	17. Bearing cone	24. Bushing
4. Bearing cone	11. Bearing cup	18. Countershaft gear	25. Reverse idler gear
5. Main drive gear	12. Shim	19. Snap ring	26. Shaft lock pin
6. Pilot bearing	13. Bearing retainer	20. Countershaft	27. Idler gear shaft
7. 1st & 3rd gear	14. Bearing retainer	21. Bearing	

end of the shaft. The torque is measured after the shaft has started to turn. The test torque is not intended to cover the force necessary to start the shaft turning. Removing shims increases bearing preload.

NOTE: If the countershaft bearing preload is on the low side of the specifications given in paragraph 115, the mainshaft bearing preload should also be on the low side. The reverse is also true.

113. MAIN DRIVE GEAR (CLUTCH SHAFT).

The main drive gear (5—Fig. FO103 or FO105) is removed from the front of the transmission housing. The gear and shaft assembly may be removed after the tractor is split between the transmission and engine as outlined in paragraph 99, without disturbing the other transmission parts. However, if bearing retainer (1—Fig. FO105), bearing (3 and 4) or gear (5) is renewed, it will be necessary to adjust the bearings, which requires removal of the transmission to gain access to the mainshaft bearing retainer shims (12) that control the adjustment.

To remove the main drive gear assembly, first drive the pin out of the clutch release fork and shaft, and remove the fork. Remove the cap screws attaching the bearing retainer (1) to the transmission housing, then withdraw the drive gear and retainer as an assembly. Pull the drive gear shaft out of the bearing retainer. Remove the bearing cup (3) and cone (4) if necessary.

NOTE: Bearing cone and cup should always be renewed as a set.

Drive or pry the oil seal (2) from the retainer. Install a new oil seal, making sure that the seal lip faces rearward (toward the transmission). Lubricate the seal with oil or grease before installing the drive gear. Assemble the main drive gear and bearings into the retainer being careful not to damage the oil seal. Reinstall the gear and retainer assembly and check the mainshaft bearing adjustment as described in paragraph 112.

114. COUNTERSHAFT.

The countershaft (20—Fig. FO103 or FO105) is removed from the rear after the mainshaft is out and before the reverse idler is removed. Remove the power take-off shifter and bearing support (32—Fig. FO104) from the rear of the transmission housing. Remove the shifter clutch hub (31) from the rear of the countershaft. Remove the front bearing retainer (14—Fig. FO105) and remove the snap ring (19) from the front of the shaft. Pull the countershaft rearward out of the housing and remove the countershaft gear (18) and front bearing cone (17).

Inspect all parts for wear or damage and renew as necessary. Installation of countershaft is reverse of removal procedure. If the countershaft or bearings

are renewed, the bearing preload must be adjusted as outlined in paragraph 115.

115. COUNTERSHAFT BEARING PRELOAD.

Reinstall the countershaft and gear in the transmission housing. Install the front bearing cone, snap ring and bearing retainer (14—Fig. FO105) with bearing cup and a new gasket. Install the pto shifter and bearing retainer assembly (32—Fig. FO104) with shims (23) as necessary to obtain zero end play of the shaft. Bearing adjustment is correct if 15-20 in.-lbs. (1.7-2.2 N·m) of torque is required to rotate the countershaft. The countershaft may be turned for checking the rolling torque by inserting the pto shaft in the shifter unit and rotating the pto shaft.

NOTE: Be sure to measure the torque after the countershaft is turning, not the torque required to start the shaft turning.

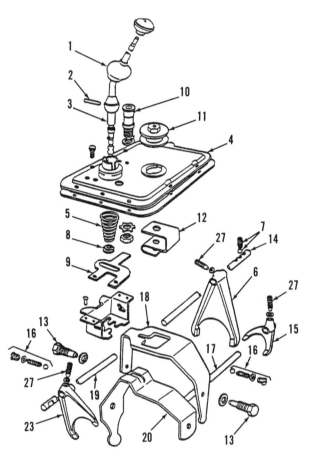

Fig. FO106—Exploded view of gear shifter mechanism on Model 8N.

1. Dust seal	12. Starter switch support
2. Pin	13. Shifter plate pivot
3. Shift lever	14. Shift rail
4. Transmission cover	15. Shift fork, reverse
5. Spring	16. Detent ball & spring
6. Shift fork, 1st & 3rd	17. Shift rail
7. Detent ball & spring	18. Gearshift plate, reverse
8. Spring seat	19. Shift rail
9. Safety start latch	20. Gearshift plate, 2nd & 4th
10. Starter switch	23. Shift fork, 2nd & 4th
11. Oil filler plug	27. Lock screws

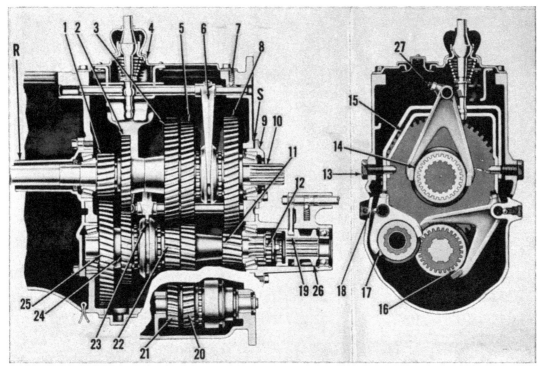

Fig. FO107—Transmission side and end views on Model 8N tractor.

R. Bearing retainer
S. Shims

1. Main drive (input) gear	7. Detent ball & spring	14. Shift coupling	21. Reverse drive gear
2. Fourth gear	8. First gear	15. Shift plate	22. Countershaft 2nd gear
3. Second gear	9. Bearing retainer	16. Shift coupling, 2nd & 4th	23. Shift fork, 2nd & 4th
4. Shift lever assy.	10. Mainshaft	17. Shift coupling, reverse	24. Countershaft 4th gear
5. Third gear	11. Countershaft	18. Detent ball & spring	25. Countershaft drive gear
6. Shift fork, 1st & 3rd	12. Pto clutch hub	19. Pto shift fork	26. Pto sliding coupling
	13. Pivot screw	20. Reverse idler gear	27. Lock screw

116. REVERSE IDLER. The reverse idler gear (25—Fig. FO105) can be removed after the countershaft is removed. Unscrew the idler shaft lock pin (26) located on outer left-hand side of the transmission housing, then withdraw the idler shaft (27) and gear (25).

Renew worn or damaged parts. Reinstall the idler gear assembly, making sure that the beveled face of the idler gear is toward the rear of the transmission.

Model 8N

117. SHIFT RAILS AND FORKS. Shift rails and forks may be removed without performing a front section split after transmission cover and gear shift lever are removed and rear axle center section is separated from the transmission case as outlined in paragraph 126.

Lift out top shifter rail spring and detent ball (7—Fig. FO106). Back off shifter fork lock screw (27) and remove shifter rail (14) and fork (6). Remove shift plate pivot screws (13) from both sides of transmission housing, then remove the shift plates (18 and 20). Remove both lower shift rails (17 and 19) and forks (15 and 23) out top opening after removing detent balls and springs (16) from outside of housing.

Renew worn or damaged parts. Reassemble by reversing the disassembly procedure. The square cornered slots in lower rails must face inward.

118. MAINSHAFT. The transmission must be removed from the tractor as outlined in paragraph 109 in order to remove the mainshaft (10—Fig. FO107 and FO108) and the gears. Remove the top shifter rail and fork, then proceed as follows: Remove the mainshaft rear bearing retainer (9) and shims (S). Remove the clutch shaft (main drive gear) and retainer (R) as a unit. Remove the reverse idler shaft shift rail, allowing the fork to remain in the housing. Lift out the mainshaft and gear cluster as a unit. Remove the bearing cones (30 and 36) from front and rear of the shaft, then slide the thrust washers (31 and 34), gears (2, 3, 5 and 8), shift coupling (32) and connector (33) off the shaft.

Renew worn or damaged parts and reassemble onto the mainshaft in the order shown in Figs. FO107 and FO108. Install the mainshaft and gears as a unit into the housing. Install the main drive gear (1) and retainer (R) assembly. Install rear baering retainer (9) and adjust mainshaft bearings as outlined in paragraph 119.

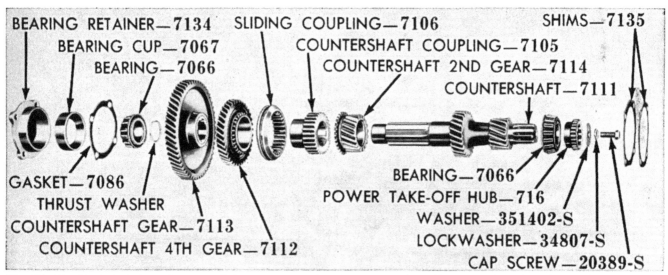

Fig. FO108—Transmission mainshaft and components on Model 8N.

R. Bearing retainer	5. Third gear	29. Bearing cup	34. Thrust washer
S. Shims	8. First gear	30. Pilot bearing	35. Bearing cup
1. Main drive gear	9. Bearing retainer	31. Thrust washer	36. Bearing cone
2. Fourth gear	10. Mainshaft	32. Coupling	37. Bearing cone
3. Second gear	28. Gasket	33. Connector	38. Oil seal

Fig. FO109—Exploded view of countershaft, gears and related components on Model 8N. The pto shifter unit is connected to the splined end of this shaft.

119. MAINSHAFT BEARING ADJUSTMENT.

If the mainshaft and/or bearings were renewed, the mainshaft bearing adjustment must be checked as follows:

Install the rear bearing retainer (9—Fig. FO107 or FO108) with original shims (S). Position the gears in neutral, then measure the torque required to rotate the mainshaft.

> NOTE: Measure the torque after the shaft is turning, not the torque necessary to start the shaft rotating from a standing start.

The bearing adjustment is correct if the rolling torque is between 20 to 35 in.-lbs. (2.3-3.9 N•m). Vary the number and thickness of shims (S) under the rear bearing retainer if adjustment is necessary.

> NOTE: If the countershaft bearing preload is on the low side of the specifications given in paragraph 122, the mainshaft bearing preload should also be on the low side. The reverse is also true.

120. MAIN DRIVE GEAR (CLUTCH SHAFT).

This gearshaft (1—Fig. FO107 and FO108) may be removed after the transmission and engine are separated (split) as outlined in paragraph 100. However, if the shaft, bearing (37) and/or retainer (R) is renewed, bearing adjustment will be required which will require removal of the transmission to gain access to the mainshaft bearing shims (S) that control the adjustment.

Unbolt and remove the retainer housing (R) and gearshaft (1) as a unit from the front of the transmission. Withdraw the gearshaft from the retainer. Remove the bearing cup (29) and cone (37) if necessary. Drive the oil seal (38) out of the retainer.

Renew worn or damaged parts as necessary. When installing a new oil seal, be sure that the lip (open side) of the seal faces rearward (toward the transmission). Lubricate the lip of the oil seal, then assemble the shaft in the retainer being careful not to damage the oil seal. After the shaft and retainer are installed,

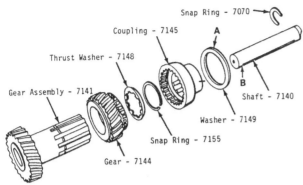

Fig. FO110—Reverse idler gear assembly on Model 8N. Some transmissions are provided with an additional washer, part No. 7029, between (A) and (B).

check the bearing adjustment. Readjust, if necessary, as outlined for the mainshaft in paragraph 119.

121. COUNTERSHAFT. This shaft (11—Fig. FO107) is removed from the rear after the mainshaft is removed, but before the reverse idler gear is removed. Remove the power take-off shifter and bearing support from the rear of the transmission. Remove the countershaft and gear assembly through the shift cover opening in the top of the transmission housing. Remove the bearing cone from the front of the countershaft, then slide the gears and shift coupling off the shaft. See Fig. FO109.

Renew worn or damaged parts. Reassemble in the reverse order of disassembly, referring to Fig. FO109. Install the countershaft and gears in the housing. Install the front bearing retainer and bearing cup with a new gasket. Adjust countershaft bearings as outlined in paragraph 122.

122. COUNTERSHAFT BEARING ADJUSTMENT. If the shaft and/or bearings were renewed, adjust the bearings by means of shims (7135—Fig. FO109) located under the pto shifter and bearing retainer housing as follows: Vary thickness of the shims as necessary to obtain zero end play of the shaft. Bearing adjustment is correct when 15-30 in.-lbs. (1.7-3.4 N•m) of torque is required to rotate the shaft. The shaft can be rotated for checking bearing adjustment by inserting the pto shaft in the shifter unit and engaging the jaw clutch.

123. REVERSE IDLER GEAR. The reverse idler gear assembly (Fig. FO110) can be removed after the countershaft and the reverse idler shift fork and rail are removed. Pull the idler shaft out toward the rear and lift out gears and shift coupling.

Renew worn or damaged parts as necessary. Reassemble in order shown in Fig. FO110.

DIFFERENTIAL, BEVEL GEARS AND REAR AXLES

DIFFERENTIAL

Models 9N-2N-8N

124. REMOVE AND REINSTALL. The differential assembly is mounted in the rear axle center housing and may be removed without disturbing the transmission or hydraulic lift as follows: Drain the lubricant from the transmission, hydraulic system and rear axle center housing. Raise and support the axle center housing, then remove the left rear wheel. Disconnect the clutch linkage (9N and 2N) and left brake linkage (8N). Disconnect the hydraulic lift lower link (3—Fig. FO111) at the left axle housing (4). Unbolt left axle housing from center housing and pull the axle housing with the shaft off the tractor. Lift the ring gear and differential assembly out of the center housing. See Fig. FO112A.

Installation of differential is the reverse of removal procedure. Be sure to use standard thickness gasket between the axle housing and rear center housing to maintain the correct differential carrier bearing setting.

125. OVERHAUL. Remove eight differential case bolts and separate the two case halves. See Fig. FO112B. Remove spider, differential pinions and

Fig. FO111—Left side view of Moel 8N axle center housing.

1. Pto control lever	3. Hitch lower link
2. Brake rod (8N)	4. Axle housing

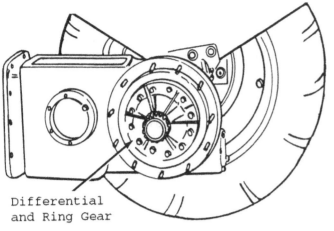

Differential
and Ring Gear

Fig. FO112A—The differential can be removed from the axle center housing after removing the left axle housing.

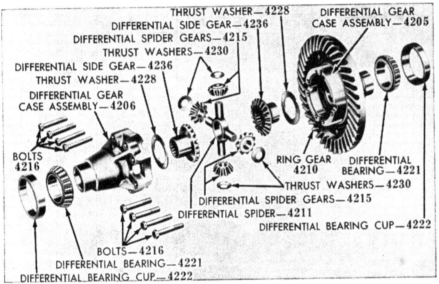

THRUST WASHER—4228
DIFFERENTIAL SIDE GEAR—4236
DIFFERENTIAL SPIDER GEARS—4215
THRUST WASHERS—4230
DIFFERENTIAL SIDE GEAR—4236
THRUST WASHER—4228
DIFFERENTIAL GEAR
CASE ASSEMBLY—4206
BOLTS
4216
DIFFERENTIAL GEAR
CASE ASSEMBLY—4205
RING GEAR 4210
DIFFERENTIAL BEARING—4221
THRUST WASHERS—4230
DIFFERENTIAL SPIDER GEARS—4215
DIFFERENTIAL SPIDER—4211
DIFFERENTIAL BEARING CUP—4222
BOLTS—4216
DIFFERENTIAL BEARING—4221
DIFFERENTIAL BEARING CUP—4222

Fig. FO112B—Exploded view of differential used in Models 9N, 2N and 8N. Bearing cups (4222) are mounted in the axle housings.

thrust washers, and lift out the side gears and thrust washers.

Carefully examine, and renew all worn or damaged parts. Differential main drive bevel gear and differential case are riveted together and are usually renewed as an assembled unit. However, if a new drive gear is installed on the old case, the case must be checked for trueness before riveting the new gear in place. After riveting is complete, check the trueness of the assembled unit. If differential carrier bearings are to be renewed, both axle housings must be off the tractor and the axle shafts removed from the housings to provide clearance for bearing cup removal and installation.

Reassembly of the differential is the reverse of disassembly procedure, plus the following: Align the matching numbers stamped on the differential case halves. Lock the case retaining bolts with wire.

TRACTOR REAR SPLIT

Models 9N-2N-8N

126. To split the tractor between the rear axle center housing and the transmission housing, first drain the oil from the housings. Remove the power take-off control (1—Fig. FO111) from the left-hand side of the center housing and cover plate from right-hand side of center housing. To assist in reassembly of the tractor, remove the cap screws attaching the pto output shaft retainer to the rear of the center housing and withdraw the pto output shaft and retainer as a unit. Refer to paragraph 139 or 140. Remove the step plates from both sides of the transmission. Disconnect the clutch linkage (9N and 2N) and left-hand brake linkage (8N). Install wooden wedges between the front axle and the axle support to prevent tipping. Block up under the axle center housing to support the rear section of the tractor. Support the front section of the tractor with a suitable rolling floor jack or overhead hoist. See Fig. FO113. Unbolt the axle center housing from the transmission and separate the tractor.

To reconnect the tractor, reverse the splitting procedure. Reinstall the pto output shaft after the transmission housing is attached to the rear axle center housing.

Fig. FO113—Rear center section separated from transmission on models 9N, 2N and 8N.

MAIN DRIVE BEVEL GEARS

Models 9N-2N-8N

127. REMOVE AND REINSTALL. To remove the main drive bevel pinion, first remove the hydraulic lift cover as follows: Place the touch control lever in the down position and move the lift arms to the fully lowered position to release trapped oil from the ram cylinder. Remove the tractor seat. Remove the pin connecting the main control spring yoke to the 3-point hitch top link lift rockers. Disconnect the hitch lift arms from the leveling arms. Remove the power take-off control plate (1—Fig. FO111) from the left side of the center housing, and working through the opening, spread the lift control fork until the lower ends of the fork are released from the pump control valve (9N and 2N models). On all tractors, remove the cap screws attaching the lift cover to the center housing and carefully lift the assembly off the center housing (Fig. FO114).

Drain the oil from the transmission and center housings. Remove the cap screws attaching the power take-off shaft retainer to the rear of the center housing, then withdraw the power take-off output shaft assembly. Refer to paragraph 139 or 140. Remove the cap screws retaining the hydraulic pump housing to the bottom of the center housing and lower the pump out of the center housing (Fig. FO114). Remove the step plates. Place wooden wedges between the front axle and axle support to prevent tipping. Support front and rear of tractor with suitable stands and a rolling floor jack or overhead hoist. Unbolt and sepa-

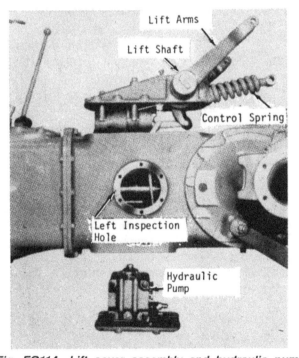

Fig. FO114—Lift cover assembly and hydraulic pump must be removed for access to main drive bevel gears.

rate the rear axle center housing from the transmission housing (Fig. FO113).

To remove the bevel pinion shaft and gear, unscrew the six cap screws (10—Fig. FO115) attaching the bearing retainer housing to the center housing, then move the bearing carrier and bevel pinion out toward the front using a suitable puller. The pinion shaft rear

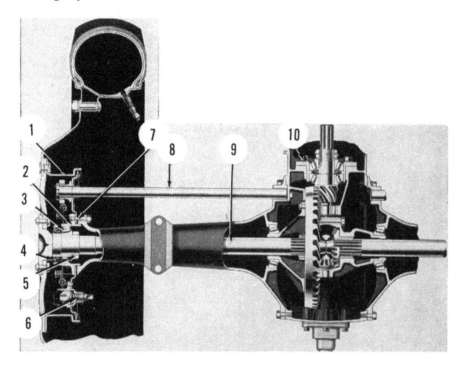

Fig. FO115—Rear axle and differential assembly as used on models 9N and 2N. The 8N is similar except details of brakes and wheel axle shafts.

1. Brake drum
2. Bearing carrier
3. Axle bearing
4. Oil seal
5. Retainer collar
6. Brake adjuster
7. Shims
8. Brake actuating camshaft
9. Axle shaft
10. Pinion carrier cap screws

bearing (located in the center housing) may be removed after the differential unit is removed.

To remove the bevel pinion ring gear, remove the left axle housing and withdraw the differential and ring gear from axle center housing (Fig. FO112A). The bevel ring gear and differential case (Fig. FO112B) are riveted together and are usually renewed as an assembled unit.

Installation of the bevel ring gear and bevel pinion is the reverse of removal procedure, plus the following: Ring gear and pinion gear mesh and backlash are not adjustable. Be sure to install a standard thickness gasket between the axle center housing and rear axle housing to maintain correct adjustment of the differential carrier bearings. Install the power take-off shaft after the axle center housing and transmission housing are reconnected and the hydraulic pump has been reinstalled. When reinstalling the lift cover, be sure that the touch control lever, implement position control lever (8N) and the lift arms are in the down position before lowering the cover into position. Working through the opening in the left side of the center housing, guide the tip of the lift control arm into position in the pump control valve lever as the cover is lowered into position on 8N models. On 9N and 2N models, reconnect the lift control fork to the pump control valve.

128. OVERHAUL. To disassemble the removed bevel pinion shaft unit, straighten the tabs on the lockwasher and remove the adjusting nuts (1—Fig. FO116) from the pinion shaft (8). Press the pinion shaft out of the bearing support (6) and front bearing cone (4). Pull the rear bearing cone (7) off the shaft

using a suitable puller. Remove the bearing cups from the retainer. Inspect and renew parts as necessary.

To remove the bevel ring gear (18), cut or drill out the rivets attaching the ring gear to the differential case (19). The case must be checked for distortion before installing the new drive gear. After riveting the new gear in place, recheck the trueness of the assembled unit.

NOTE: It is recommended that the bevel pinion gear and ring gear be renewed as a set. Installing only one new gear may result in noisy operation and accelerated wear of gear teeth.

To reassemble the bevel pinion, press the rear bearing cone onto the pinion shaft. Press the bearing cups into the retainer housing. Insert the pinion shaft into the retainer and install the front bearing cone and adjusting nuts. Adjust the preload of the pinion bearings by turning the adjusting nuts as shown in Fig. FO117 until a torque of 12-16 in.-lbs. (1.4-1.8 N•m) is required to turn the shaft.

NOTE: Be sure to measure the torque after the shaft is rotating, not the torque required to start the shaft turning from a standing start.

After the bearing adjustment is correct, bend the tabs on the lockwasher to secure the nuts.

AXLES AND BEARINGS

Models 9N-2N

129. REMOVE AND REINSTALL. Either rear axle shaft (9—Fig. FO115) can be removed without removing axle housing from tractor if so desired. To

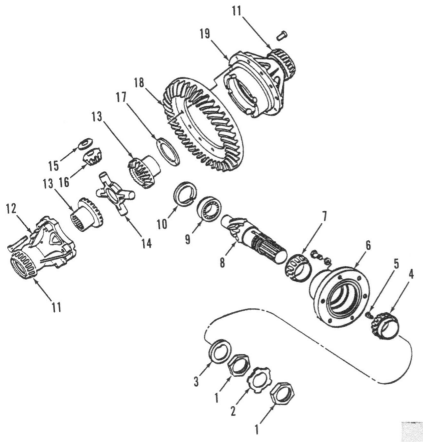

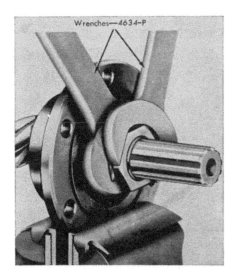

Fig. FO116—Exploded view of differential assembly and main drive bevel gears.

1. Nuts
2. Lockwasher
3. Thrust washer
4. Bearing
5. Pin
6. Bearing carrier
7. Bearing
8. Bevel pinion shaft
9. Pilot bearing
10. Lock ring
11. Bearings
12. Differential case half
13. Differential side gears
14. Differential spider
15. Thrust washer
16. Differential spider gear
17. Thrust washer
18. Ring gear
19. Differential case half

Fig. FO118—Removing axle shaft and brake assembly on Models 9N and 2N. The shims (A) adjust end play of wheel axle shafts.

18. Axle shaft	
19. Axle housing	27. Brake back plate
26. Brake drum	29. Brake shaft

Fig. FO117—Adjusting the bearings of main drive bevel pinion.

remove either rear axle, first raise and support the tractor. Remove the rear wheel and fender support brace. Remove the stud nuts attaching the axle bearing retainer to the axle housing. See Fig. FO118. Pull the axle shaft and brake as an assembly from the axle housing. Pack axle bearing with wheel bearing grease prior to installation. To install the axle assembly, reverse the removal procedure. Adjust the axle bearings as outlined in paragraph 131.

130. RENEW AXLE BEARINGS. With axle removed from tractor, turn brake adjuster eccentric counterclockwise to back the brale shoes away from brake drum, then lift off the brake and backing plate assembly to expose the axle bearing retainer collar (5—Fig. FO115). The retainer collar is a shrink fit on the axle shaft and can be removed by drilling a ¼ inch (6 mm) hole through the retainer collar, then split the collar at the drilled hole as shown in Fig. FO119. A

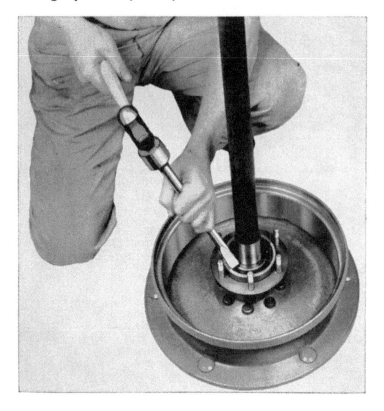

Fig. FO119—Removing the shrink fit wheel axle shaft bearing retainer collar.

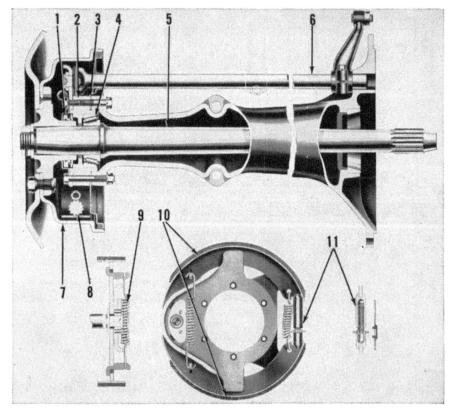

Fig. FO120—Left axle shaft and housing assembly on early Model 8N. Later 8N models are similar except they include an inner oil seal (not shown) for each axle.

1. Oil seal
2. Bearing retainer
3. Shims
4. Axle bearing
5. Axle shaft
6. Brake shaft
7. Brake drum
8. Shoe adjuster wheel
9. Shoe retracting spring
10. Brake shoes
11. Shoe adjusting screw

cutting torch may also be used to cut the shrink fit sleeve, but be careful not to damage the axle. Remove the bearing carrier (2—Fig. FO115), bearing (3) and collar from the axle shaft using a suitable puller. Remove axle seal (4) from bearing carrier.

Reassemble parts in reverse order of disassembly, using a new oil seal (4), bearing (3), and retainer collar (5). Lubricate bearing and lip of seal with grease prior to installation of axle. The retainer collar must be heated to expand it, then pressed or driven

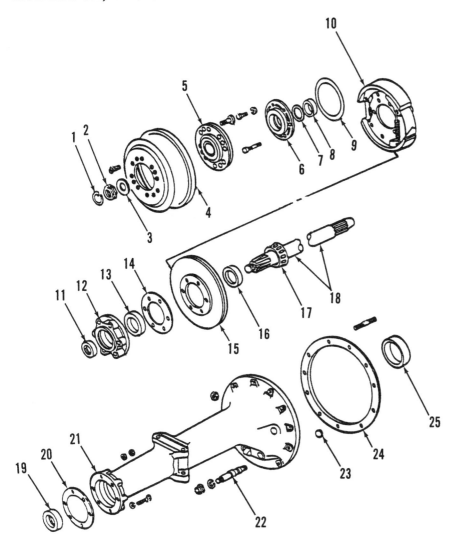

Fig. FO121—Exploded view of Model 8N rear axle assembly. Oil seal retainer (6), gasket (7) and oil seal (8) were used on tractors prior to serial number 486753.

1. Snap ring
2. Nut
3. Washer
4. Brake drum
5. Wheel hub
6. Oil seal retainer
7. Gasket
8. Oil seal, outer
9. Gasket
10. Brake shoe assy.
11. Oil seal, outer
12. Bearing retainer
13. Bearing cup
14. Shim
15. Brake backing plate
16. Cork seal
17. Bearing cone
18. Axle shaft
19. Oil seal, inner
20. Shim
21. Axle housing
22. Lower link support shaft
23. Expansion plug
24. Gasket
25. Bearing cup

onto the axle shaft until it bottoms against the inner race of the axle bearing.

131. ADJUST AXLE BEARINGS. To check axle shaft bearing adjustment, jack up rear of tractor and remove rear wheels. Rotate either axle shaft and observe whether the opposite shaft rotates in the same or opposite direction. If both shafts rotate in the same direction, bearings are adjusted too tightly. Check for excessive axle shaft end play by attempting to move axle in and out. If any motion can be detected, bearings are adjusted too loosely.

To adjust axle bearings, remove shims (A—Fig. FO118) from between either the right or left bearing retainer and axle housing until both shafts rotate in the same direction when one is turned. Then add shims on either side until shafts start turning in opposite directions. This procedure will provide end play of 0.002-0.006 inch (0.05-0.15 mm). Shims are available in three different thicknesses.

Model 8N

132. REMOVE AND REINSTALL. Either rear axle shaft (5—Fig. FO120) can be removed without removing axle housing from tractor if desired. To remove either rear axle, first raise and support the rear of the tractor. Remove the rear wheel.

NOTE: If axle is to be disassembled, it is recommended that snap ring (1–Fig. FO121) and axle retaining nut (2) be removed prior to removing axle assembly from tractor as the nut is tightened to a very high torque and may be difficult to remove.

Turn brake adjuster screw (Fig. FO123) to back the brake shoes away from brake drum. Remove the brake drum retaining screws, then remove the brake drum (7—Fig. FO120) from the wheel hub. Remove the cap screws and stud nuts attaching the bearing retainer (2) to the axle housing. Pull the axle (5) and brake assembly from the axle housing.

Pack axle bearing with wheel bearing grease prior to installation. To install axle, reverse the removal procedure. Adjust the axle shaft bearings as outlined in paragraph 134. Tighten axle shaft nut (1—Fig. FO121) to 450 ft.-lbs. (610 N·m) torque and install lock ring (1).

133. OVERHAUL. To disassemble the axle components, remove the lock ring (1—Fig. FO121) and wheel hub retaining nut (2) if not previously removed. Use a suitable puller to remove the wheel hub (5) from the axle shaft (18). Withdraw the axle shaft from the bearing retainer (12).

Axle shaft bearing (17) and oil seals can be renewed at this time. Be sure that cork oil seal (16) is installed against the bearing cone (17).

Reassemble parts in reverse order of disassembly. Pack axle bearing with wheel bearing grease. Tighten the axle shaft nut (2) to a torque of 450 ft.-lbs. (610 N·m) and install the lock ring (1).

134. ADJUST AXLE BEARINGS. To check axle shaft bearing adjustment, jack up rear of tractor and remove rear wheels. With transmission in neutral, rotate either axle shaft and observe whether the opposite shaft rotates in the same or opposite direction. If both shafts rotate in the same direction, bearings are adjusted too tightly. Check for excessive axle shaft end play by attempting to move axle in and out. If any motion can be detected, bearings are adjusted too loosely.

To adjust axle bearings, remove shims (3—Fig. FO120) from between either the right or left bearing retainer and axle housing until both axle shafts rotate in the same direction when one is turned. Then add shims on either side until axle shafts start turning in opposite directions. This procedure will provide end play of 0.002-0.006 inch (0.05-0.15 mm). Shims are available in various thicknesses.

BRAKES

Models 9N-2N

135. ADJUSTMENT. To adjust the brakes, jack up rear of tractor until the rear wheels are free to turn. Turn brake adjusting stem (Fig. FO122) clockwise until the wheel can no longer be turned by hand. Then, back off adjusting stem until rear wheel turns with a very slight brake drag. Adjust both sides equally.

Model 8N

136. ADJUSTMENT. To adjust the brakes, jack up rear of tractor until the rear wheels are free to turn. Remove the cover from the adjusting screw opening (Fig. FO123) in brake back plate. Turn notched adjuster screw until the wheel can no longer be turned

by hand. Then, back off the adjuster until rear wheel turns with a very slight drag. Adjust both sides equally. Adjust left brake rod clevis (2—Fig. FO111) to equalize the brake pedals when both brakes are applied.

BRAKE SHOES

Models 9N-2N

137. REMOVE AND REINSTALL. Jack up rear end of tractor and remove the rear wheels and fender support brace. Remove the stud nuts attaching the axle bearing retainer to the axle housing. Pull the axle shaft and brake as an assembly out of the axle housing (Fig. FO118). Turn brake adjuster eccentric (Fig. FO122) counterclockwise to back the brake

Fig. FO122—Brake adjuster stem on Models 9N and 2N.

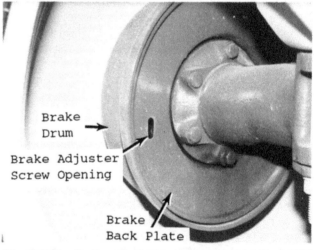

Fig. FO123—Brake adjuster screw opening on Model 8N.

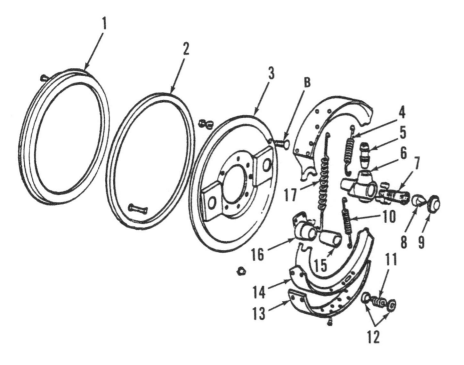

Fig. FO124—Exploded view of brakes and related parts used on 9N and 2N models.

1. Dust shield ring
2. Dust seal
3. Brake backing plate
4. Return spring (green)
5. Brake shoe link
6. Brake adjuster bracket
7. Brake adjuster stem
8. Wedge
9. Cap
10. Return spring (red)
11. Hold down spring
12. Spring cups
13. Brake lining
14. Brake shoe
15. Bushing
16. Brake camshaft bracket
17. Return spring (blue)

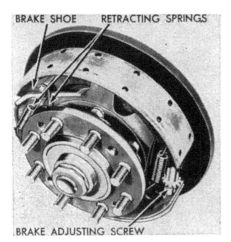

Fig. FO125—Model 8N rear wheel hub and brake assembly.

shoes away from brake drum. Lift the brake backing plate (3—Fig. FO124) off the bearing retainer studs. Remove brake shoe retracting springs (4, 10 and 17) and hold down springs (11). Remove brake shoes (14).

The brake lining (13) is riveted to the brake shoe, and new lining is available for installation on a used brake shoe. New brake shoe and lining are also available as an assembled unit.

Reinstall brake shoes in reverse order of removal. Be sure that upper brake shoe fits squarely in slot of adjuster link (5—Fig. 124). If necessary, adjust equalizer bolt (B—Fig. 124) to align the shoe correctly. Note that upper brake shoe return spring (4) is color coded

green and lower spring (10) is color coded red. Adjust the brakes as outlined in paragraph 135.

Model 8N

138. REMOVE AND REINSTALL. Jack up rear of tractor and remove the rear wheels. Turn brake adjuster screw (Fig. FO123) to back the brake shoes away from brake drum. Remove the four retaining screws from the brake drum, then pull the drum off the axle. Remove the brake shoe retracting springs. See Fig. FO125. Remove the shoes, adjusting screw and spring as a unit.

NOTE: Right-hand brake uses only one retracting spring (9—Fig. FO126). Left-hand brake uses two retracting springs (7 and 9).

Brake shoes and lining assemblies are available in a kit for one wheel only. (Two kits are required to renew both brakes.) One shoe and lining fits either right-hand upper or left-hand lower position; the other shoe fits either right-hand lower or left-hand upper position.

To reinstall brake shoes, hook adjusting spring (11—Fig. FO126) into rear end of shoes and place end of shoes into notches of adjusting screw assembly (12, 13 and 14). Spread front end of shoes and install on anchor plate. Reinstall brake retracting springs (only one retracting spring used on right-hand brake) and spring anchors (10). Adjust brakes as outlined in paragraph 136.

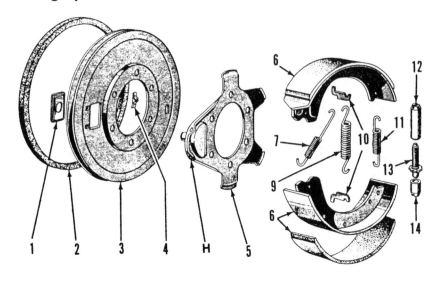

Fig. FO126—Exploded view of 8N brake assembly. Spring anchors (10) fit in slots in brake shoes, and brake retracting spring (7) hooks into upper spring anchor and hole (H) in anchor plate (5). Spring (7) is not used on right brake. Same brake backing plate (3) and anchor plate (5) are used on both right and left brake assemblies. Left brake assembly is shown; right brake is inverted.

1. Shaft hole cover	9. Retracting spring
2. Felt dust seal	10. Spring anchors
3. Backing plate	11. Adjusting screw
4. Adjusting hole cover	spring
5. Anchor plate	12. Nut
6. Brake shoe assy.	13. Adjusting screw
7. Retracting spring	14. Socket

POWER TAKE-OFF UNIT

A transmission driven type power take-off is available on all models. The front end of the pto shaft is supported in a ball bearing mounted in the pto shifter housing which is attached to the rear wall of the transmission housing. The rear end of the pto shaft is supported in a ball bearing mounted in a retainer on the rear face of the axle center housing. On all models, the pto shaft rotates at 545 rpm when engine speed is at 1500 rpm. Pto engagement and disengagement is controlled by hand lever (1—Fig. FO111) on left side of axle center housing.

OUTPUT SHAFT

Models 9N-Early 2N

139. REMOVE AND REINSTALL. To remove the pto output shaft (1—Fig. FO127), first drain the oil from rear axle center housing, hydraulic system and transmission (three plugs). Remove the cap screws attaching the rear bearing retainer (7) to the axle center housing. Withdraw the shaft and retainer assembly.

NOTE: A pto conversion kit is available from various sources that sell tractor parts and accessories. The conversion kit replaces the original pto output shaft with a new 1 3/8 inch splined shaft. The kit is a ready to install assembly that includes shaft, bearing, bearing retainer, seal, cover and cap.

To renew rear oil seal assembly, remove the seal case (13), spring (2), seal ring (11) and bearing seal (9). To renew bearing (6), the shrink fit bearing retainer sleeve (5) must be split using a sharp chisel.

Sleeve may also be removed using a cutting torch if care is taken not to damage pto shaft. The bearing and sleeve can then be pulled from pto shaft. The new sleeve must be heated until it turns blue in color to facilitate installation and insure a tight shrink fit.

Installation of pto shaft assembly is reverse of removal procedure. Tighten bearing retainer mounting bolts to 53-60 ft.-lbs. (72-81 N·m) torque.

Model 8N-Late 2N

140. REMOVE AND REINSTALL. To remove the pto output shaft, first drain the transmission and hydraulic system fluid on 8N and 2N models (three plugs). Remove the cap screws attaching the rear bearing retainer (3—Fig. FO128) to the axle center housing, and withdraw the pto shaft assembly.

NOTE: A pto conversion kit is available from various sources that sell tractor parts and accessories. The conversion kit replaces the original pto output shaft with a new 1 3/8 inch splined shaft. The kit is a ready to install assembly that includes shaft, bearing, bearing retainer, seal, cover and cap.

Rear oil seal (5) may be renewed as follows: Remove front snap ring (8) and pull bearing retainer (3) from bearing and shaft. Remove the rear snap ring (6) and press the seal from retainer. Reverse the removal procedure to install a new seal. Inspect sleeve (4) and remove any burrs or rough spots before reassembling.

Rear bearing (7) may be renewed while retainer (3) is removed from the pto shaft. Split the shrink-fit sleeve (4) with a sharp cold chisel. The sleeve may also be removed using a cutting torch if care is taken

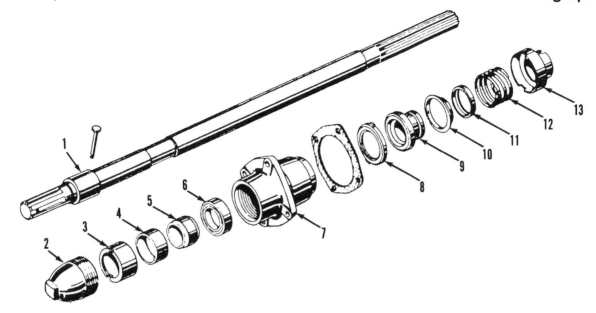

Fig. FO127—Power take-off shaft assembly as used on Model 9N and early production 2N tractors.

1. Pto output shaft
2. Shaft cover cap
3. Bearing retainer
4. Spacer
5. Sleeve
6. Bearing
7. Bearing retainer
8. Seal case washer
9. Bearing seal
10. Spring guide
11. Seal ring
12. Spring
13. Seal case

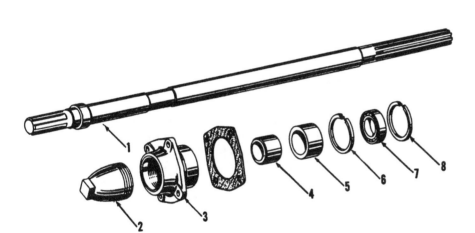

Fig. FO128—Components of power take-off shaft assembly typical of 8N and late production 2N tractors. The bearing retainer sleeve (4) is a shrink fit on the shaft.

1. Shaft & sleeve
2. Cap
3. Bearing retainer
4. Sleeve
5. Oil seal
6. Snap ring
7. Bearing
8. Snap ring

not to damage pto shaft. Pull the bearing and sleeve from pto shaft. Press new bearing on shaft. Heat new sleeve evenly until color turns to blue, then drop sleeve on shaft against bearing. After sleeve is cool, remove any scale or rough spots to avoid damage to seal (5).

Reverse the removal procedure to reinstall the pto shaft assembly. Tighten bearing reatiner mounting bolts to 53-60 ft.-lbs. (72-81 N•m) torque.

PTO SHIFTER UNIT

All Models

141. R&R AND OVERHAUL. To remove the pto shifter unit and front bearing support (Fig. FO129),

it will be necessary to split the tractor between the transmission and axle center housing as follows: Drain the lubricant from transmission and center housing. (Three plugs on all models.) Remove shifter plate (1—Fig. FO111) from left side of center housing. Remove both foot step plates. Disconnect clutch linkage on 9N and 2N and left brake linkage (2) on 8N. Place wood wedges between the front axle and axle support to prevent tipping. Support transmission housing and axle center housing using suitable safety stands and a floor jack. Unbolt center housing from the transmission, then separate the tractor. See Fig. FO113.

Remove the cap screws attaching the shifter unit to rear wall of transmission housing and withdraw the shifter unit. Be careful not to lose the shims

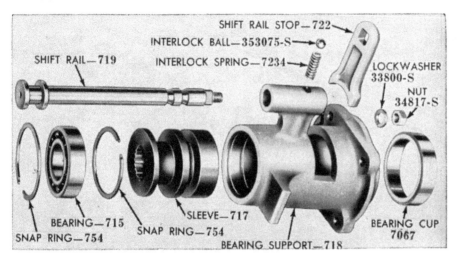

Fig. FO129—Power take-off shifter assembly and pto shaft front bearing support as used on Models 9N, 2N and 8N. The housing also functions as the transmission countershaft rear bearing support.

located between the support and transmission housing.

To remove the bearing, remove outer snap ring and pull the bearing from the support. See Fig. FO129. To remove the shifter sleeve, first remove the nut from shift rail and withdraw the rail from the shifter fork and support housing. Be careful not to lose the detent ball and spring. Remove the snap ring from support and withdraw shifter sleeve.

Reassemble the shifter unit and install to rear of transmission with same number of shims as were removed. Tighten retaining cap screws to a torque of 25-30 ft.-lbs. (34-40 N·m).

BELT PULLEY

The belt pulley is supplied as optional equipment and may be mounted and operated in right or left horizontal position or in down vertical position. The belt pulley revolves at 1358 rpm when the crankshaft speed is 2000 rpm, or 1494 rpm when the crankshaft speed is 2200 rpm.

All Models So Equipped

142. R&R AND OVERHAUL. Removal of the belt pulley unit requires the removal of four cap screws which hold the unit to rear axle center housing. To install the unit remove the pto cap and the four cap screws which retain the lower link swing chains. Engage the splines on the pto shaft. Locate the pulley in the desired operating position and secure with the four cap screws.

Drain lubricant and remove housing cover (13—Fig. FO130) and gasket. Remove castellated nut (6) from inner end of pulley shaft. This nut must be unscrewed in stages as the shaft is being removed. Withdraw shaft (1) and remove drive shaft and gear (11) out through housing cover opening.

Pulley shaft gear (5) and drive shaft gear (11) are furnished only as a matched set. The mesh and backlash of these gears is fixed and nonadjustable. Install oil seals (2 and 9) with lips facing inward. Adjust pulley shaft bearings to a just perceptible preload by means of pulley nut (6) prior to installing cover (13). Bearing preload is correct when 5-12 in.-lbs. (0.6-1.3 N·m) torque is required to rotate the shaft in its bearings. Adjust the drive shaft bearings to a slight preload by varying the number of gaskets installed between cover (13) and housing (7). Bearing preload is correct when 15-34 in.-lbs. (1.7-3.8 N·m) torque is required to rotate the belt pulley when unit is completely assembled. Shim gaskets for drive shaft bearings are available in various thicknesses.

Belt pulley housing holds one pint (0.47 L) of lubricant, or level with filler plug opening. Use SAE 90 EP gear lubricant.

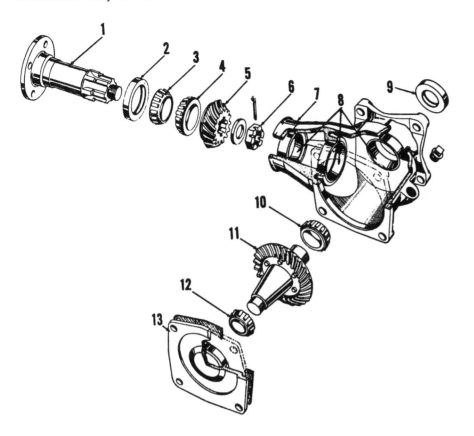

Fig. FO130—Components of the optional equipment belt pulley assembly.

1. Pulley shaft
2. Oil seal
3. Shaft bearing cone
4. Shaft bearing cone
5. Pulley shaft gear
6. Gear retaining nut
7. Housing
8. Bearing cups
9. Oil seal
10. Drive shaft bearing
11. Drive gear & shaft
12. Drive shaft bearing
13. Cover & gasket

HYDRAULIC SYSTEM

HYDRAULIC LIFT OPERATION

All Models

143. Although all of the 9N, 2N and 8N tractors use main assemblies of basically similar appearance, the hydraulic system as used on the 9N and 2N tractors performs differently and has different adjustments than the system used on the 8N. Main differences are in the valving, which in the 9N and 2N provide automatic draft control, whereas in the 8N, the system has automatic draft control plus automatic implement control (sometimes called automatic follow-up control). Each system is easily identified by the appearance of the lift cover assembly, which, if it is for an 8N tractor, has a "position control" lever (Fig. FO131) and a touch control lever, whereas for the 9N and 2N units, the cover has only the touch control lever.

Pressure fluid to operate the hydraulic lift system is supplied by the hydraulic pump located in the bottom of the rear axle center housing and driven by the tractor pto shaft. See Fig. FO131. The pto must be engaged to operate the hydraulic lift mechanism. The hydraulic pump housing contains a combined intake and exhaust control valve on Models 9N and 2N that controls pump output and is actuated by the lift control linkage. The Model 8N has separate valves for intake and exhaust which are connected by a rocker (Fig. FO132).

On all models, the Scotch Yoke pump runs continuously whenever the tractor and pto shaft are in use, but when the control valves are in neutral or lowering position the fluid cannot flow. The outlet side of the Scotch Yoke pump is an open line to the ram cylinder and contains the relief valve which also serves as a safety valve in the event of shock loads on the lift.

When the intake control valve is in the raising position (valve open), fluid going through the valve can enter the inlet valve and pass through the Scotch Yoke pump to the ram cylinder. When the intake valve closes, fluid can no longer flow to the Scotch Yoke pump which continues to operate but pumps no fluid because the supply is cut off by the "neutral" control valve. When the exhaust control valve moves to the "exhaust" or lowering position, the fluid trapped in the ram cylinder can flow through the control valve and back into the reservoir, but the intake portion of the control valve remains closed so no fluid is supplied to the Scotch Yoke pump and it continues to idle, operating, but not pumping fluid.

144. RAISE LIFT ARMS. To raise the lift arms, the control linkage (Fig. FO132) moves the intake control valve to open the pump intake port, permit-

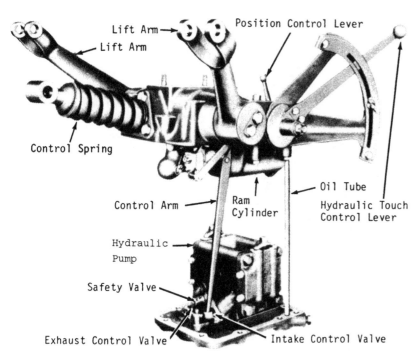

Fig. FO131—View of hydraulic pump and lift cover as used on 8N tractors. Models 9N and 2N are similar except that position control lever is not used.

ting fluid to flow into the Scotch Yoke pump where it is automatically fed to the ram cylinder. The lift continues to raise until flow through the intake port is cut off, at which time oil remains static in the Scotch Yoke pump and in the ram cylinder. The safety valve is in the static pressure fluid, and continues to protect the lift from shock loads.

145. LOWER LIFT ARMS. To lower the lift arms, the control linkage (Fig. FO132) moves the exhaust valve to open the pump exhaust port and fluid flows out the port from the lift cylinder. The intake portion of the control valve remains closed so no fluid is fed to the Scotch Yoke pump which continues to idle. The lift arms will continue to lower until the control valve moves back to neutral or until all fluid is exhausted from the ram cylinder. If a controlled lowering is

signaled, the exhaust valve will move back to the neutral position to stop the action.

146. HOLD LIFT ARMS IN POSITION. When the intake control valve and exhaust control valve are both in neutral position (Fig. FO132), intake and exhaust ports are closed. Operating fluid is trapped in the ram cylinder and the passage leading to the safety valve, which continues to serve as a protection factor. The Scotch Yoke pump continues to idle because fluid is prevented from flowing to it by the closed intake valve.

Incidental lowering because of leakage from the ram cylinder circuit will be compensated for by intermittent movement of the intake valve to admit a momentary flow of fluid followed by a return to the

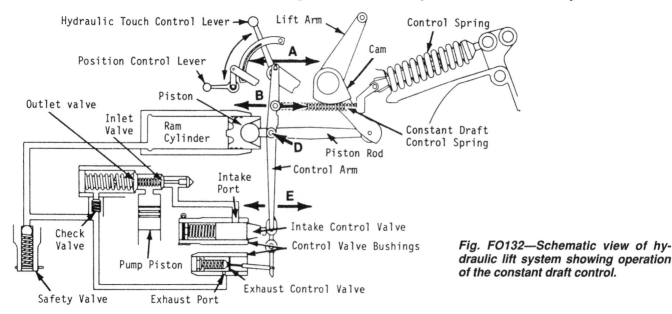

Fig. FO132—Schematic view of hydraulic lift system showing operation of the constant draft control.

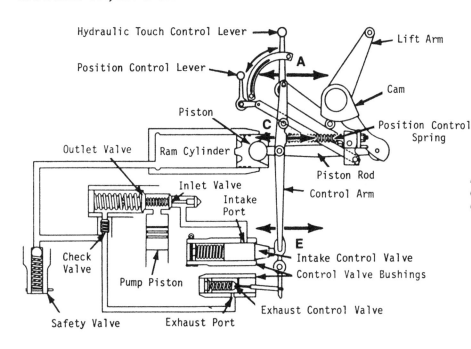

Fig. FO133—Schematic view of hydraulic lift system showing operation of position control (8N).

neutral position. Unlike a true bypass system, there can be no "hiccuping" because the slight opening of the intake valve will feather the flow of fluid to the system and the change is hardly noticed.

147. CONSTANT DRAFT CONTROL. For constant draft control operation, the position control lever (8N) is placed in the fully forward position. See Fig. FO132. The lift control arm pivots at points A, B, D, and E. Pivot point "A" is manually positioned by the setting of the touch control lever. Pivot point "B" is positioned by the implement draft load through the control spring. Point "D" is controlled by the ram piston. Point "E" is mechanically connected to the control valves.

When the touch control lever is moved forward, point "A" moves forward, point "B" is stationary and point "E" moves rearward opening the exhaust control valve port. Fluid is discharged from the ram cylinder, the implement enters the soil and the draft load compresses the control spring. As the spring compresses, points "B" and "E" are moved forward. When the desired draft setting is reached, point "E" will be in the neutral position, closing the exhaust port. The implement will remain at that depth until the draft load increases or decreases, or the touch control lever is moved.

When the draft load increases, such as when the front wheels cross a ridge, forcing the implement downward into the soil, the increased draft will move points "B" and "E" forward, opening the intake port. Fluid is then pumped to the ram cylinder to raise the implement and automatically maintain the draft load setting.

The opposite reaction occurs when the draft load decreases, such as when the front wheels enter a depression which tends to raise the implement out of the soil. The decrease in draft load will allow the control spring to expand, which moves points "B" and "E" rearward. The exhaust port is opened and fluid is discharged from the ram cylinder allowing the implement to lower. As the implement lowers and draft load increases to the preselected setting, the control spring is compressed and points "B" and "E" move forward allowing the exhaust control valve to return to neutral position.

When the touch control lever is moved rearward, point "A" moves rearward, point "B" is stationary and point "E" moves forward. This opens the pump intake port and fluid is pumped to the ram cylinder, raising the lift arms. As the implement raises, the draft load decreases and the control spring expands, causing points "B" and "E" to move rearward. The implement will continue to raise until point "E" reaches the neutral position and the intake port closes.

When the touch control lever is moved to the top of the quadrant for transport, point "A" is moved rearward, point "B" is stationary and point "E" moves forward. The pump intake port is opened and fluid is pumped to the ram cylinder which raises the lift arms. When the lift arms reach the fully raised position, the skirt on the ram piston contacts point "D" and moves "D" and "E" rearward until control valves are in neutral position.

148. POSITION CONTROL (8N). For position control operation, the position control lever is placed in the vertical position. See Fig. FO133. The hydraulic lift control arm then pivots at points "A", "C" and

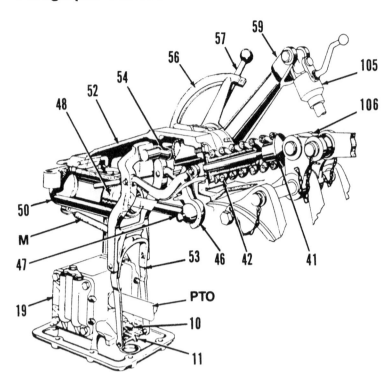

Fig. FO134—Phantom view of hydraulic lift cover and pump as used on Models 9N and 2N.

M. Fork retracting spring
10. Pump relief valve
11. Control valve
19. Pump unit
41. Control spring yoke
42. Main control spring
46. Lift ram arm
47. Connecting rod
48. Ram piston
50. Ram cylinder
52. Lift cover
53. Control lever fork
54. Control lever shaft
56. Quadrant
57. Touch control lever
59. Lift shaft arm
105. Leveling box
106. Lift rocker

"E". Pivot point "A" is moved by manual operation of the touch control lever. Pivot point "C" is moved by the cam on the lift ram arm. Point "E" is mechanically connected to the control valves.

When the touch control lever is moved forward, point "A" is moved forward, point "C" is stationary and point "E" is moved rearward, which opens the pump exhaust port and closes the intake port. Fluid is discharged from the ram cylinder through the exhaust port and the lift arms lower. As the lift arms lower, the cam on the lift ram arm moves points "C" and "E" rearward until the control valves are in neutral position. The lift arms will remain in this position until the touch control lever is moved again.

When the touch control lever is moved rearward, point "A" is moved rearward, point "C" is stationary and point "E" is moved forward, which opens the pump intake port and closes the exhaust port. Fluid is pumped to the ram cylinder and the lift arms raise until the lift arm cam allows points "C" and "E" to move rearward, placing the control valves in the neutral position.

ADJUSTMENTS

Models 9N-2N

149. QUADRANT. To check the quadrant adjustment, place an implement or a weight of 200-300 pounds (90-135 kg) on the lift arms and move the touch control lever (57—Fig. FO134) upward to fully raise the lift arms (59). Then, move the touch control lever downward slowly while observing the lift arms.

The lift arms should start to lower when the touch control lever reaches a point 2⅝-2¾ inches (66.7-69.8 mm) down from the top position. If lift arms start moving before or after the specified lever position, loosen the four quadrant attaching cap screws and move quadrant (56) toward front or rear as required.

If touch control lever will not remain in any set position, tighten the castellated nut at external end of touch control lever shaft. Recommended frictional drag is when a pull of 4-5 pounds (1.8-2.2 kg) at end of lever is required to move the lever.

150. CONTROL SPRING. To check the control spring adjustment, place lift in UP position and remove the pin which secures the control spring yoke (41—Fig. FO134) to the lift rocker (106). Using a thumb-to-finger grip, note if spring (42) can be rotated. If spring cannot be rotated or if it has any end play, adjust by turning the threaded yoke. Adjustment is correct when spring is free to turn, but no end play exists.

ADJUSTMENTS

Model 8N

151. QUADRANT. To check the quadrant adjustment, first remove the inspection plate (X—Fig. FO135) from the right side of the axle center housing. Disengage the position control lever (62) (smaller lever of the two) on the lift cover by moving it forward to the horizontal position. Insert left hand through inspection plate opening and hold the hitch control

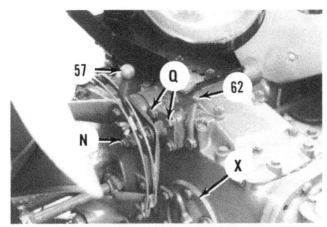

Fig. FO135—Model 8N hydraulic system has automatic draft control and position control. The position control function is not used on 9N and 2N tractors.

N. Friction adjusting nut
Q. Quadrant unit screws (4)
X. Inspection plate
57. Touch control lever
62. Position control lever

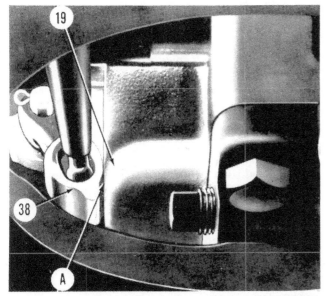

Fig. FO136—On Model 8N, the quadrant is correctly positioned when rocker lever (38) first contacts housing (19) just as the touch control lever reaches its full UP position on the quadrant.

rocker lever (38—Fig. FO136). While slowly moving the touch control lever (57—Fig. FO135) toward top of quadrant with the right hand, determine by sight and feel when the end of intake valve is flush with the pump housing (rocker lever contacts pump housing).

If the quadrant position is correct, valve rocker lever (38—Fig. FO136) will first contact the pump housing (19) at point (A), just as the touch control lever reaches its full UP position on the quadrant. If valve lever does NOT contact the pump housing simultaneously with the touch control lever arriving at the full UP position, loosen the cap screws (Q—Fig. FO135) in the quadrant support plate and move the

Fig. FO137—Turn nut (N) to set touch control lever frictional drag.

Fig. FO138—Adjust main control spring (42) by rotating the threaded yoke (41) until all end play of spring is removed.

quadrant forward or backward as required. Top face of quadrant support plate should be parallel with top of the attaching plate on the lift cover after adjustment is completed.

If touch control lever will not remain in any set position, tighten the castellated nut (N—Fig. FO137) at external end of touch control lever shaft. Recommended frictional drag is when a pull of 4-5 pounds (1.8-2.2 kg) at end of lever is required to move the lever.

152. MAIN CONTROL SPRING. To check the main control spring adjustment, place the lift in UP position and remove the pin (P—Fig. FO138) to disconnect yoke (41) from lift rocker. Note if the spring (42) can be rotated by hand using moderate force. If the spring cannot be rotated (too tight) or if it has any end play (too loose), adjust by rotating the threaded yoke as necessary. Adjustment is correct when spring can be turned, but no end play exists.

NOTE: The following adjustments described for the constant draft spring and position control spring should be required only when the lift cover assembly has been disassembled. The adjustments

Fig. FO139—On Model 8N, adjust the constant draft control spring to 3-9/16 inches (90.5 mm) length by means of adjusting nut (81).

LP. Pad surface of
    control arm
R. Marks indicating top
    position of lift arms
62. Position (drawbar)
    control lever
68. Position (drawbar)
    control lever

78. Constant draft control spring
79. Shoulder flange on draft
    control bushing
80. Swivel
81. Nut for adjusting length
    of draft control spring

can be made only when the lift cover is off the tractor.

153. CONSTANT DRAFT SPRING. To adjust the constant draft spring, mount the lift cover assembly in a vise with control spring up as shown in Fig FO139. Disengage the position control lever (62) by moving it to the forward (down) position. Measure the length of the constant draft control spring (78) which should be $3\frac{9}{16}$ inches (90.5 mm) plus or minus $\frac{1}{64}$ inch (0.4 mm). See Fig. FO140. If spring is not within these limits, adjust to $3\frac{9}{16}$ inches (90.5 mm) by means of adjusting nut (81—Fig. FO139).

154. POSITION CONTROL SPRING. To adjust the position control spring, mount lift cover assembly in a vise as shown in Fig. FO141 and engage the position control lever (62) by moving it to the UP position. Move the touch control lever down until there is a gap "D" of $\frac{3}{4}$ inch (19 mm) between edge of lever and top of slot as shown. Raise the lift arms to their top operating position as indicated by marks (R) being in register. Be sure the control arm (84) moves freely. Move position control spring linkage up until pin (71) is in contact with face of cam on lift ram arm. Raise the control arm (84) until swivel (80—Fig. FO140) comes into contact with the flanged shoulder on bushing (79) of the constant draft control spring. With parts held in this position, loosen the locking

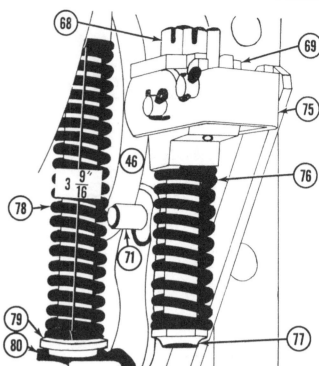

Fig. FO140—Drawing of control springs and position of linkage in hydraulic lift cover on 8N tractors.

46. Cover
68. Locknut for position
    control spring
69. Cam plate
71. Pin for position control arm
75. Position control arm

76. Position control spring
77. Position control rod
78. Constant draft spring
79. Draft control bushing flange
80. Swivel

nut (68) and turn the control rod (77) located inside spring (76) until the bolt contacts the pad (LP—Fig. FO139) on the control arm. Recheck the position of touch control lever and lift arms to be sure they were not moved during the adjustment, then tighten locknut (68—Fig. FO140) securely.

After the lift cover is reinstalled on the tractor, a final check of the position control spring adjustment can be made as follows: Move the implement position control lever rearward to the UP position. Move touch control lever to bottom of the quadrant. When lift arms are fully lowered and the tractor drawbar installed, measure the distance from the rear end of the lower links to the ground. Then move the position control lever fully forward and check the amount of drop of the lift arms by remeasuring the distance from the end of the lower links to the ground. If the drop exceeds 3 inches (76 mm), the position control spring length is too long.

TROUBLESHOOTING

155. The hydraulic system on the 9N, 2N, or 8N tractors is easily examined and troubles isolated. When troubleshooting the hydraulic system, always check the simplest and most obvious items first. Much time can be saved in servicing the hydraulic

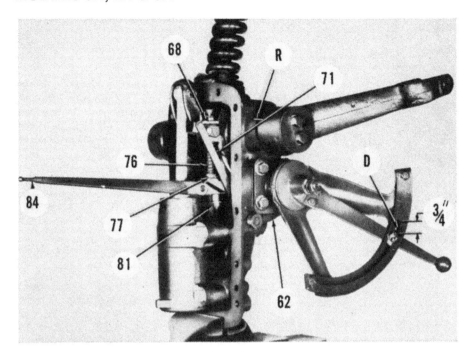

Fig. FO141—Adjusting the position (drawbar) control spring (76) on 8N tractor.

  R. Marks indicating top position
      of lift arms
62. Position (drawbar) control lever
68. Locknut for position control spring
71. Pin
76. Position control spring
77. Lower end of position control rod
81. Nut for adjusting length of
      constant draft spring
84. Lift control arm

system by correctly locating the cause or causes of any difficulty before attempting repairs. Some of the problems which may occur while operating the hydraulic system and their possible causes are as follows:

Hitch Will Not Lift
1. Low fluid level
2. Faulty pump
3. Relief valve not seating
4. Sticking control valve
5. Cracked cylinder or piston

Will Not Hold Load in Lift Position
1. Scored ram cylinder
2. Faulty piston or seal ring
3. Faulty safety valve
4. Check valve leaking

Noisy Pump Operation
1. Low fluid level
2. Air in hydraulic system
3. Faulty pump

ERRATIC CONTROL. Erratic control is usually caused by leaks or sticking valves in the pump. Leaks can usually be detected by careful observation with the inspection cover removed from right side of axle center housing. Sticking valves can only be determined by disassembly of the pump as outlined in paragraph 162 (8N) or 169 (9N, 2N).

Erratic control may also be the result of binding control linkage or a faulty check valve.

ERRATIC LIFTING. Damage to the ram piston can cause a condition where the lift will seem to handle a normal load but will fail to lift immediately after attaching a load to the lift arms, where the lift arms must be manually raised to connect the load.

The condition occurs when the piston connecting rod (dog bone) punches the closed end out of the piston, then seals the hole by contact with the piston. To check the condition, disconnect the lift arms from the load, move position control lever and touch control lever rearward to "raise" position and recheck lift action. If the lifting action now appears normal, remove inspection cover from right side of the center housing and check for oil flow from the ram cylinder area when lift arms are held away from the ram piston.

ERRATIC DEPTH CONTROL. If the tillage implement alternates between too deep and too shallow, probable causes are main control spring not properly adjusted (paragraph 150 or 152) and/or binding linkage.

IMPLEMENT CANNOT BE LOWERED TO FULL WORKING DEPTH. Possible cause of this condition is incorrect adjustment of quadrant. Refer to paragraph 149 or 151.

FAILS TO LOWER. Possible causes of this condition is either the exhaust control valve is stuck in closed position, exhaust oil passage is plugged or the mechanical control linkage is broken or binding. Remove inspection cover from right side of center housing and check for proper movement of the control linkage and exhaust valve. If inspection does not reveal the problem, remove hydraulic unit as outlined in paragraph 161 and inspect for plugged oil passage.

FAILS TO LIFT. The most frequent causes for failure to lift are having the pto disengaged or the auxiliary transmission (SHERMAN) in neutral. With these two potential problems checked, all other problems can be checked by removing the inspection cover on the right side of the center housing and visually examining for faulty or erratic operation of any of the

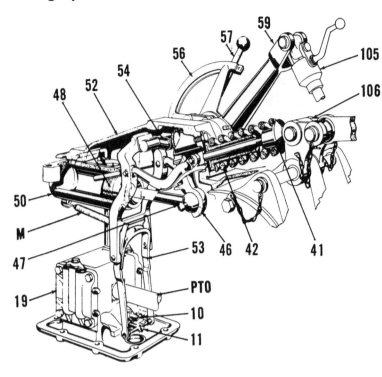

Fig. FO142—Phantom view of hydraulic lift cover (52) and pump (19) used on 9N and 2N tractors. Spread the control lever fork (53) slightly to disconnect it from the control valve (11) prior to removing the lift cover or pump (19).

M. Fork retracting spring	50. Ram cylinder
10. Pump relief valve	52. Lift cover
11. Control valve	53. Control lever fork
19. Pump unit	54. Control lever shaft
41. Control spring yoke	56. Quadrant
42. Main control spring	57. Touch control lever
46. Lift ram arm	59. Lift shaft arm
47. Connecting rod	105. Leveling box
48. Ram piston	106. Lift rocker

components. The presence of operating fluid is immediately obvious. If the hydraulic pump is submerged, there is no lack of fluid. Start the tractor and check to see that the pump is turning, if it is not, check for broken parts in the pto drive, transmission or clutch. If the pump is turning, check to be sure that control valves move when quadrant is actuated. If control valves move to the raising position and pump is turning, fluid is being pumped and must be going somewhere. Visually check for oil flow or leakage in pump valving, ram cylinder or the supply tube pressed into the center housing. Renew any obviously defective parts or overhaul the system as outlined in paragraph 157 or 159.

LIFT COVER

Models 9N-2N

156. REMOVE AND REINSTALL. To remove the hydraulic lift cover, first fully lower the lift arms to release trapped fluid from the lift cylinder. Remove the tractor seat. Remove the pin attaching the main control spring yoke (41—Fig. FO142) to lift rocker (106). Disconnect leveling rods (105) from lift arms (59). Remove pto shifter plate from left side of axle center housing (Fig. FO143) and working through the inspection cover opening, spread the lift control fork (53—Fig. FO142) until lower ends of fork are released from control valve (11). It is not necessary to remove the hydraulic pump. Remove the cap screws securing lift cover to rear axle center housing and carefully lift cover off center housing.

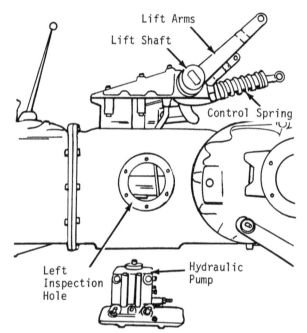

Fig. FO143—Lift cover assembly and hydraulic pump being removed from rear axle center housing.

When reinstalling the lift cover, reverse the removal procedure using new gaskets, and reconnect the lift control fork to control valve. Tighten lift cover retaining bolts to 53-60 ft.-lbs. (72-81 N•m) torque. Adjust the quadrant and main control spring as outlined in paragraphs 149 and 150.

157. DISASSEMBLY AND OVERHAUL. To disassemble the lift cover components, unbolt and remove the ram cylinder (50—Fig. FO144) with piston (48). Unscrew the control spring yoke (41) and remove

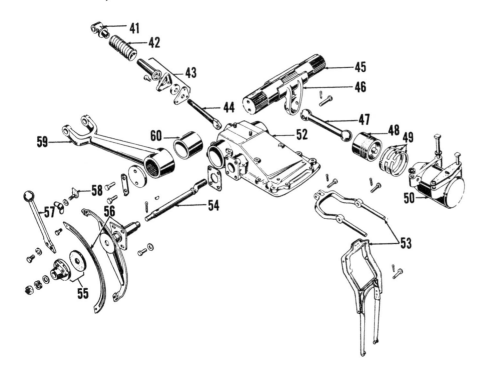

Fig. FO144—Exploded view of 9N and 2N hydraulic lift cover assembly.

41. Yoke
42. Control spring
43. Plunger guide
44. Plunger
45. Lift shaft
46. Ram arm
47. Connecting rod
48. Ram piston
49. Piston rings
50. Ram cylinder
52. Lift cover
53. Fork
54. Shaft
55. Friction plate
56. Quadrant
57. Lever
58. Quadrant stop
59. Arm
60. Bushing

main control spring (42) and plunger assembly (43 and 44). Disconnect and remove the lift control lever fork (53). Remove the cap screws retaining the lift arms (59) to the lift shaft (45) and remove the lift arms. Bump the end of the lift shaft either way to remove it along with one of the bushings (60) from the lift cover (52) and ram arm (46). Push the other bushing out of the lift cover. Unbolt and remove hand control quadrant (56) and related parts.

Clean all parts with a suitable solvent. Renew any worn, damaged or corroded parts. Clearance of piston (48) in ram cylinder (50) should not exceed 0.0025 inch (0.063 mm). Bushings (60) are presized and will require no final sizing if carefully installed. Check linkage for being worn, bent or twisted and renew as necessary.

Reassemble the unit by reversing the disassembly procedure. Be sure to lubricate the piston seal rings (49) with oil, and install piston carefully in ram cylinder to avoid damaging the seal rings. Tighten ram cylinder mounting bolts to 53-60 ft.-lbs. (72-81 N•m) torque. When installing the lift arms, be sure notch marks on arms and ends of lift shaft are aligned. Tighten the lift arm retaining cap screws until the lift shaft binds, then loosen the screws until arm can be raised and lowered freely. Bend tabs of lock plates to secure the lift arm cap screws.

Model 8N

158. REMOVE AND REINSTALL. To remove the lift cover, first move the touch control lever and the position control lever to down position to lower the lift arms and release trapped fluid from the lift cylinder. Remove the tractor seat. Remove the pin attach-

ing main control spring yoke to lift rocker. Disconnect lift arms from leveling arms. Remove the cap screws securing lift cover to center housing and lift off the cover (Fig. FO143).

Installation of the lift cover is the reverse of removal, plus the following: Be sure that the touch control lever, implement position control lever and lift arms are in the down position before lowering the cover into position. Remove the inspection cover from the right side of the axle center housing; then working through the opening, guide the tip of the control arm into position in the control valve rocker lever as the cover is lowered into position on the center housing. See Fig. FO145. After lift cover is in position, move touch control lever to determine whether the control valves follow the lever movement without binding. Tighten lift cover retaining bolts to 53-60 ft.-lbs. (72-81 N•m) torque. Adjust the quadrant and main control spring as outlined in paragraphs 151 and 152.

159. DISASSEMBLY AND OVERHAUL. Unscrew the control spring yoke (41—Fig. FO146) and remove the control spring (42) and spring guide assembly (43). Unbolt and remove ram cylinder (50—Fig. FO147). Remove nut (81) and cotter pin, then withdraw draft control swivel (80), bushing (79) and spring (78) from contstant draft control link (73). Remove nut (68), plate (69), position control spring (76) and control rod (77). Disconnect and remove position control link (65), control arm (70) and control cam (75). Remove nut (83) attaching control lever (84) to control lever shaft (54—Fig. FO146). Remove four cap screws attaching control quadrant (56) to lift cover and withdraw control quadrant and control

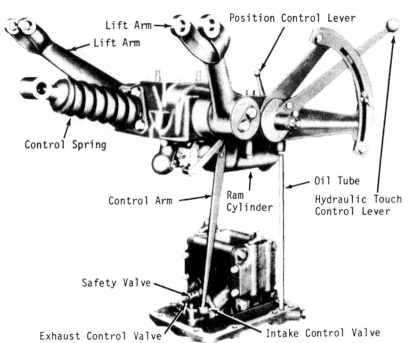

Fig. FO145—View of hydraulic pump and lift cover as used on 8N tractors.

lever shaft (54). Remove locking nut and spring (101), and separate control lever shaft (54) from friction plate (55) and touch control lever (57). Remove control spring plunger (44) and draft control link (73—Fig. FO147). Remove cap screws and retainer plates attaching lift arms (59) to lift shaft (45). Scribe match marks on the left arms and ends of the lift shaft to ensure correct alignment when reassembling, then remove the lift arms. Bump lift shaft (45) out of lift cover and remove lift ram arm (46).

Clean all parts with a suitable solvent. Renew parts that are worn, damaged or corroded. Clearance of piston in ram cylinder should not exceed 0.0025 inch (0.064 mm). New lift shaft bushings (60) are presized and should require no final sizing if carefully installed. Check all linkage for bent or twisted parts.

Reinstall the lift shaft, ram arm and lift arms in the lift cover as shown in Fig. FO148. Tighten the lift arm retaining screws until the lift shaft binds, then loosen the screws until the arms can be raised and lowered freely. Bend the tabs of the lock plates (Y—Fig. FO149) to secure the cap screws (Z). Position the draft control link (73) and main control spring plunger in the cover, then assemble the control linkage and quadrant on the cover in reverse order of disassembly procedure as shown in Fig. FO150. Be sure to lubricate piston seal rings, and install the piston carefully in the ram cylinder with new gasket to avoid damaging the seal rings. Install the ram cylinder on the lift cover as shown in Fig. FO151. Tighten ram cylinder mounting bolts to 53-60 ft.lbs. (72-81N·m) torque. Adjust the constant draft spring and the position control spring as outlined in paragraphs 152 and 153.

HYDRAULIC PUMP R&R AND TEST

Models 9N-2N-8N

160. PRESSURE TEST. To check the pump operating pressure with pump installed in the tractor, proceed as follows: Remove the hexagon head cap plug (14—Fig. FO152) located in right hand corner of the pump base, and in its place connect a 0-2000 psi (0-14,000 kPa) pressure gauge.

> NOTE: Transmission/hydraulic fluid should be at normal operating temperature when checking pump pressure.

With the lift arms secured in the fully lowered position or weighted to prevent them from raising, start the engine and move the touch control lever to the top (raising) position on the quadrant. With engine running at about 1500 rpm, the gauge pressure should be 1500-1700 psi (10,345-11,720 kPa).

If the pump does not deliver the specified minimum pressure, check for a faulty safety (relief) valve (30—Fig. FO152) as follows: Remove either the inspection hole cover or the pto shift lever cover from the right or left side of rear axle center housing. There should be visible fluid turbulence in the vicinity of the safety valve. If turbulence is not evident and pump pressure is less than 1500 psi (10,345 kPa), the safety valve is probably not at fault. Other points of leakage are also indicated by turbulence or leakage when viewed through the inspection port. Leakage is also sometimes indicated by noises or heating of the operating

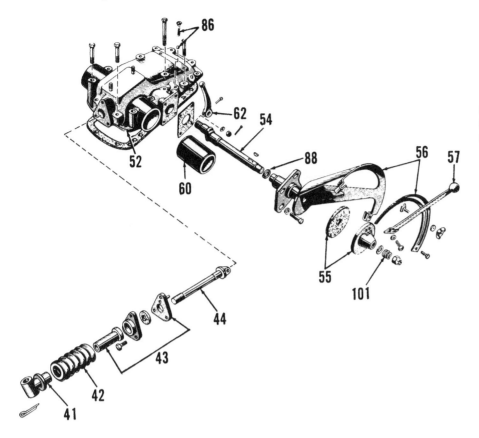

Fig. FO146—Model 8N lift housing cover (52), touch control lever (57), control spring (43) and related parts.

41. Yoke
42. Main control spring
43. Guide assy.
44. Plunger
52. Cover
54. Control lever shaft
55. Friction plate & disc
56. Quadrant
57. Touch control lever
60. Bushing
62. Position control lever
86. Detent
88. Washer
101. Spring

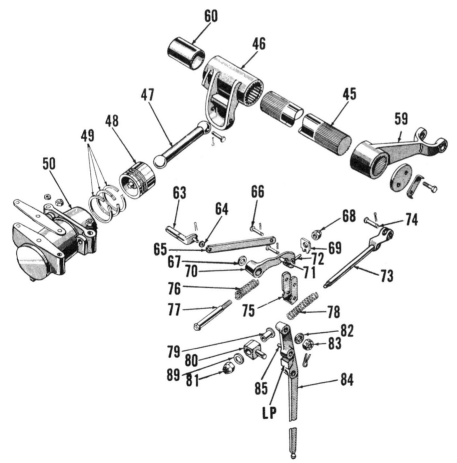

Fig. FO147—Exploded view of Model 8N ram cylinder (50), lift shaft arms (59), constant draft control rod (73), position control lever and linkage which are mounted on the hydraulic lift cover.

45. Lift shaft	71. Dowel pin
46. Ram arm	72. Dowel pin
47. Connecting rod	73. Draft control link
48. Ram piston	74. Clevis pin
49. Piston rings	75. Control cam
50. Ram cylinder	76. Position control
59. Lift arm	spring
60. Bushing	77. Control rod
63. Position control	78. Draft control
arm	spring
64. Washer	79. Bushing
65. Control link	80. Control swivel
66. Clevis pin	81. Adjusting nut
67. Washer	83. Nut
68. Locknut	84. Control lever
69. Plate	85. Dowel pin
70. Control arm	89. Washer

fluid. If fluid spray, heating or noises are not evident, a further check involving removal of the pump will be necessary.

161. REMOVE AND REINSTALL. To remove the pump, first drain the fluid from the rear axle center housing. Fully lower the lift to release all trapped

Fig. FO148—Model 8N lift cover showing first step in reassembly procedure.

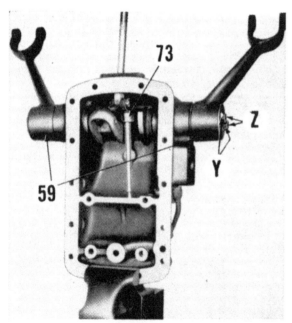

Fig. FO149—Model 8N lift cover showing second step in reassembly procedure.

Y. Lock plate   59. Lift shaft arms
Z. Cap screws   73. Control link

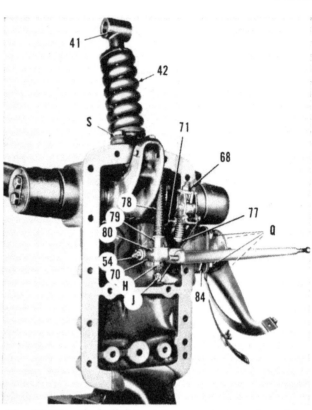

Fig. FO150—Model 8N housing cover when viewed from lower side with ram cylinder removed.

H. Control spring   70. Control arm
J. Cotter pin   71. Pin
Q. Cap screws   77. Control rod
41. Yoke   78. Draft control spring
42. Control spring   79. Bushing
54. Control lever shaft   80. Swivel
68. Locknut   84. Control lever

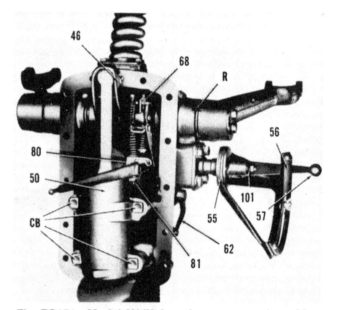

Fig. FO151—Model 8N lift housing cover as viewed from the lower side.

CB. Cylinder bolts   62. Control lever
46. Lift ram arm   68. Locknut
50. Ram cylinder   80. Swivel
55. Friction plate & disc   81. Adjusting nut
56. Quadrant   101. Spring

Fig. FO152—Model 8N hydraulic pump assembly showing the intake control valve (29), safety relief valve and check valve unit (30) and rocker lever (38) for actuation of the control valves. The exhaust control valve is shown at (34). Valve chamber covers (C) are also the pump cylinders.

fluid. Remove the cap screws attaching the pto bearing retainer to the rear of the center housing and withdraw the pto shaft. Remove the inspection port cover and pto shift lever cover from right and left sides of the center housing. On 9N and 2N tractors, disconnect the fork (53—Fig. FO142) from the control valve (11) by separating the fork to free the valve. On all models, remove the cap screws retaining the pump housing to the bottom of the center housing and lower the pump assembly from the tractor.

When reinstalling the pump, reach through the inspection holes and make certain that the control lever fork (9N and 2N) or control arm (8N) properly engages the pump control valve arm. Install the pump retaining cap screws finger tight, then install the pto shaft. Rotate the engine several revolutions with pto engaged to permit pump to center itself, then tighten the retaining screws to 53-60 ft.-lbs. (72-81 N.m) torque.

PUMP OVERHAUL

Model 8N

162. DISASSEMBLE AND OVERHAUL. To disassemble the hydraulic lift pump, refer to Fig. FO153 and proceed as follows: Remove the relief valve (30) and check valve (32). Remove the valve rocker lever (38) while holding the intake control valve (29) in against the spring pressure. Remove the intake con-

trol valve and exhaust control valve (34) from the pump housing. Remove both valve chambers (C) and the cam and piston assembly (15, 16 and 17). Remove the retaining screw (1) and clamp (2) from the valve chambers and remove the inlet and outlet valve components from the chambers.

Thoroughly clean all parts with a suitable solvent. Renew any worn, corroded or damaged parts. Check and service components of the pump as follows.

163. INTAKE CONTROL VALVE. The intake control valve (29—Fig. FO153) must have a minimum clearance fit in the valve bushing (24), but must also move freely in the bushing. When checking the fit, the valve and bushing should be coated with clean hydraulic fluid. Renew the valve and/or bushing to obtain the desired fit.

Intake valve was originally available in three different diameters. Valve diameters and corresponding color code are as follows:

Part No.	Diameter	Color
8N698E	0.5917-0.5918 in.	White
8N698F	0.5918-0.5920 in.	Blue
8N698G	0.5920-0.5921 in.	Yellow

The valve bushing is available in standard bore size only. A bushing with a 0.010 inch (0.25 mm) oversize outside diameter was originally available, but is no longer serviced by Ford. When renewing the bushing, be sure to use the latest style bushing which has four intake holes.

Position the bushing so the cotter pin hole is horizontal to facilitate insertion of the locking cotter pin. Press the bushing in flush with the face of the pump base. Inside diameter of the bushing should not be sized after installation.

164. EXHAUST CONTROL VALVE. The exhaust control valve (34—Fig. FO153) and its bushing (B) should be free from scratches or other damage. Valve should move freely in bushing, but should have minimum clearance when checked with clean transmission fluid on the valve and bushing. Exhaust valve was originally available in five different diameters to provide a means of select fitting the valve to the bushing. The valve diameters and corresponding color code are as follows:

Part No.	Diameter	Color
8N640A2	0.5917-0.5918 in.	White
8N640B	0.5919-0.5920 in.	Blue
8N640C	0.5920-0.5921 in.	Yellow
8N640D	0.5923-0.5924 in.	Green
8N604E	0.5925-0.5926 in.	Orange

Bushing for exhaust control valve is pressed into pump base and should not be scratched or otherwise damaged. Bore diameter of bushing should be 0.5926 inch (15.052 mm). Bushing should be flush with face

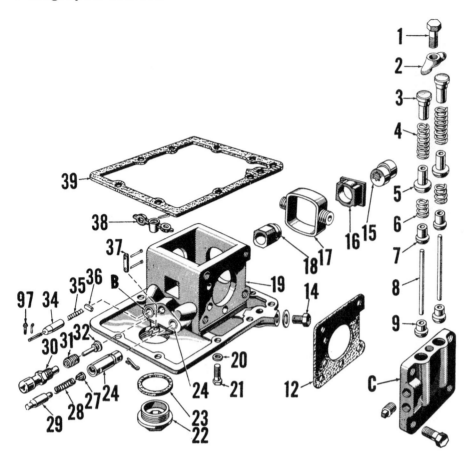

Fig. FO153—Exploded view of 8N hydraulic lift pump unit.

- B. Bushing
- C. Valve chamber
- 1. Cap screw
- 2. Valve clamp
- 3. Plug
- 4. Spring
- 5. Outlet valve
- 6. Valve spring
- 7. Inlet valve
- 8. Valve guide
- 9. Socket
- 12. Gasket
- 14. Plug
- 15. Pump cam
- 16. Cam block
- 17. Piston (2 used)
- 18. Pto shaft bushing
- 19. Pump base
- 22. Drain plug
- 23. Gasket
- 24. Bushing
- 27. Button plug
- 28. Valve spring
- 29. Intake control valve
- 30. Relief valve
- 31. Valve spring
- 32. Check valve
- 34. Exhaust control valve
- 35. Valve spring
- 36. Retainer pin
- 37. Pivot pin
- 38. Rocker lever
- 39. Gasket
- 97. Clevis pin

of pump base and with holes in such position as to assure easy insertion of the locking cotter pin.

165. SAFETY AND CHECK VALVE UNIT. A check valve (32—Fig. FO153) is combined with the safety (relief) valve (30) mounted in the pump base. The safety valve is factory adjusted to unseat at a minimum pressure of 1500-1700 psi (10,345-11,720 kPa). If any parts of this valve are worn or broken or show signs of leakage, install a new valve unit.

Inspect the check valve for signs of leakage. Improper seating may be due to irregularities of the seat in the pump base which can be corrected by using the special Ford reseating tool or equivalent as shown in Fig. FO154.

166. INLET AND OUTLET VALVES. Inlet and outlet valves (5 and 7—Fig. FO153) are contained in each of the valve chamber covers. If inspection indicates valves are not seating properly, they can be reseated using special Ford reseating tool or equivalent as shown in Fig. FO155 and the valves can be refaced. If valves leak after this work has been accomplished, make sure that guides (8—Fig. FO153) and sockets (9) are aligned to give concentric seating and that they are in good mechanical condition.

167. PTO SHAFT BUSHING. Bushing (18—Fig. FO153) which supports the pto shaft should have a running clearance of not less than 0.0015 inch (0.038 mm) and not more than 0.006 inch (0.152 mm). Renew the bushing if scored or if clearance exceeds 0.006 inch (0.152 mm).

168. SCOTCH YOKES AND CYLINDERS. The Scotch Yokes (17—Fig. FO153), which form the pistons, and the large bores in the chamber covers (C), which form the cylinders of the pump, should be inspected for scratches, wear, or other damage. Correction of any nonstandard condition is by renewing the parts.

When assembling yokes (pistons), position them so that the piston portions are closest to each other.

Models 9N-2N

169. DISASSEMBLE AND OVERHAUL. This pump differs from the unit used on the 8N tractors mainly in that it is equipped with a single, two ended control valve (11—Fig. FO156) instead of an inlet and exhaust valve such as is used on the 8N. The single control valve has a T-shaped attaching connection and is operated by the control fork, whereas on the 8N pump the control valves are operated by a rocker

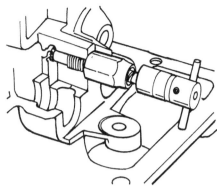

Fig. FO154—Improper seating of the check valve portion of relief valve unit may be corrected by using a reseating tool as shown.

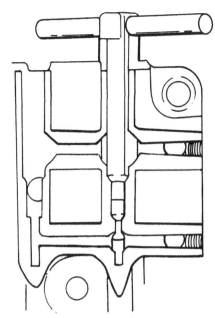

Fig. FO155—Hydraulic pump inlet and outlet valve seats in the pump valve chamber can be reseated as shown.

lever which is in turn operated by a single control lever.

To disassemble the pump, refer to Fig. FO156 and proceed as follows: Remove the safety relief valve (10) and the control valve (11). Remove both valve chambers (C) and the cam and piston assemblies (15, 16, and 17). Remove the retaining screw (1) and clamp (2) and disassemble inlet and outlet valve components (3-9) from the valve chambers (C).

Thoroughly clean all parts with a suitable solvent. Renew any worn, corroded or scratched parts. Check and service the components as outlined below.

170. PUMP CONTROL VALVE. The pump control valve (11—Fig. FO156) is available in only one diameter. It should be a minimum clearance fit in the bushing (B), but still move freely. The bushing is available with standard inside diameter only.

171. SAFETY AND CHECK VALVE. A check valve (CV—Fig. FO156) is combined with the safety (relief) valve (RV). If inspection shows that check valve has been leaking, check for irregularities of the seat surface in the pump base. Seat can be recut by using special Ford reseating tool or equivalent as shown in Fig. FO154. If seat is satisfactory but leakage still occurs, the valve should be renewed, followed by light lapping if necessary.

The safety valve portion of the assembly indicated by the spring (RV), ball and cotter pin is factory adjusted to unseat at a pressure of 1500-1700 psi (10,345-11,720 kPa). If any parts of this valve are worn or broken or show leakage, install a new valve unit.

172. INLET AND OUTLET VALVES. Inlet and outlet valves (5 and 7—Fig. FO156) are contained in each of the valve chamber covers. If inspection indicates valves are not seating perfectly, they can be reseated using the special Ford reseating tool or equivalent as shown in Fig. FO155 and the valves can be refaced. If valves leak after this work has been accomplished, make sure that guides (8—Fig. FO156) and sockets (9) are aligned to give concentric seating and that they are in good mechanical condition.

173. PTO SHAFT BUSHING. Bushing (18—Fig. FO156) which supports the pto shaft should have a running clearance of not less than 0.0015 inch (0.038 mm) and not more than 0.006 inch (0.152 mm). Renew the bushing if scored or if clearance exceeds 0.006 inch (0.152 mm).

174. SCOTCH YOKES AND CYLINDERS. The Scotch Yokes (17—Fig. FO156), which form the pistons, and the large bores in the chamber covers (C), which form the cylinders, should be inspected for wear, scratching or other damage. Correct any nonstandard condition by renewing the parts.

When assembling yokes (pistons to pump), position them so that the piston portions are closest to each other.

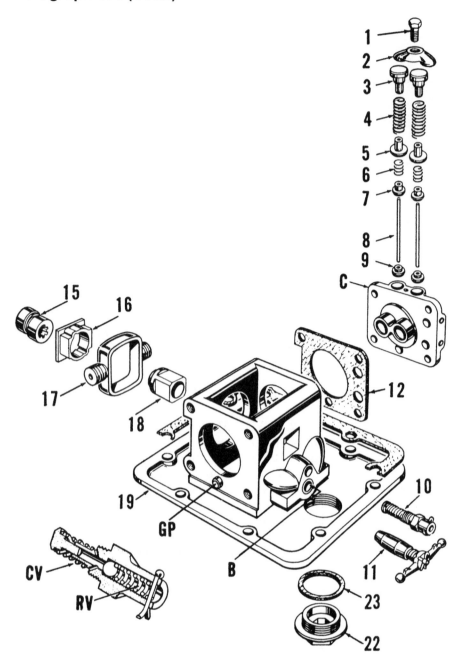

Fig. FO156—Exploded view of 9N and 2N hydraulic lift pump unit and related parts.

   C. Valve chamber
  CV. Check valve
  GP. Guide pin
  RV. Valve spring
   1. Cap screw
   2. Valve clamp
   3. Plug
   4. Valve spring
   5. Outlet valve
   6. valve spring
   7. Inlet valve
   8. Valve guide
   9. Socket
  10. Relief valve
  11. Control valve
  12. Gasket
  15. Pump cam
  16. Cam block
  17. Pump piston (2 used)
  18. Pto shaft bushing
  19. Pump base
  22. Drain plug
  23. Gasket

PART NUMBERS

175. This section contains a complete parts breakdown and listing of part numbers for the Models 9N, 2N and 8N tractors. The listing is in response to many requests for information on part numbers. It is solely for the convenience of users of the manual to enable them to correctly identify and locate the parts that they need. Note that part numbers are not shown for many of the standard fasteners, as these items are available from many sources.

The part numbers listed are the original equipment Ford numbers. Many of the original equipment parts are still available from the manufacturer and can be purchased from a New Holland Ford tractor dealer. However, the manufacturer no longer services some of the part numbers listed. Therefore, other sources for these parts may need to be located.

Some alternative sources for parts are aftermarket parts supply companies, tractor salvage yards and tractor parts rebuilders.

QUICK REFERENCE INDEX

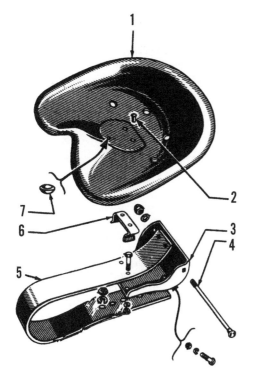

Fig 157—Operator's seat and related parts.
1. Seat 9N400C (9N, 2N) 8N400A (8N)
2. Screw 61794S
3. Hinge 8N436A (8N)
4. Bolt 3/8 x 7-1/2
5. Spring 9N404 (9N, 2N) 8N404A (8N)
6. Clip 8N433
7. Bumper 8N434B

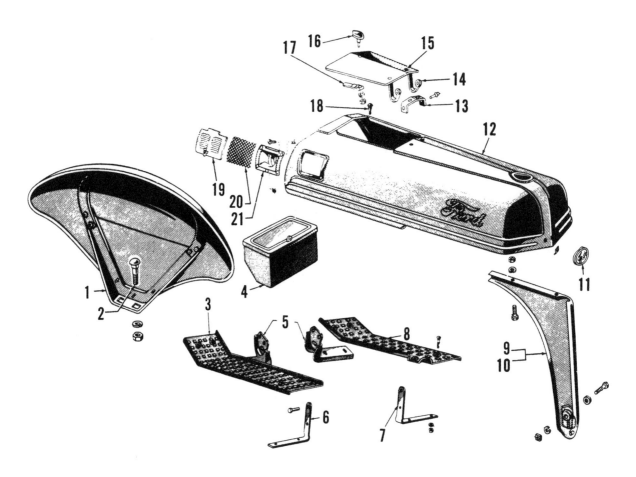

Fig. 158—Hood, fenders and related parts.

1. Fender 9N16312
    (9N, 2N) 8N16312 (8N)
2. Bolt 5/8 x 6
3. Step 8N16473
4. Tool box 8N17005
5. Bracket 8N16470
6. Support 8N16487
7. Support 8N16486
8. Step 8N16472
9. Panel RH 8N16618A
10. Panel LH 8N16619
11. Emblem 2N16600
    (9N, 2N) 8N16600A (8N)
12. Hood 9N16612C
    (9N, 2N) 8N16612 (8N)
13. Retainer 9N16651
14. Hinge 9N16689
15. Cover 9N16938
16. Stud 9N16625
17. Tongue 9N16656
18. Bumper 11A16758A
19. Cover 8N9661
20. Screen 8N9669
21. Funnel 8N9684

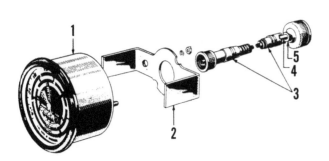

Fig. 159—Proof-meter and related parts.

1. Proof-meter 8N17360A
2. Bracket 8N17368B
3. Cable assy. 8N17365
4. Shaft 8N17366
5. Retainer 8N17372

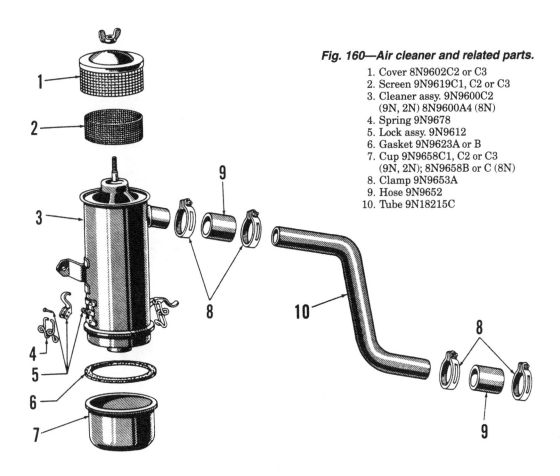

Fig. 160—Air cleaner and related parts.

1. Cover 8N9602C2 or C3
2. Screen 9N9619C1, C2 or C3
3. Cleaner assy. 9N9600C2
   (9N, 2N) 8N9600A4 (8N)
4. Spring 9N9678
5. Lock assy. 9N9612
6. Gasket 9N9623A or B
7. Cup 9N9658C1, C2 or C3
   (9N, 2N); 8N9658B or C (8N)
8. Clamp 9N9653A
9. Hose 9N9652
10. Tube 9N18215C

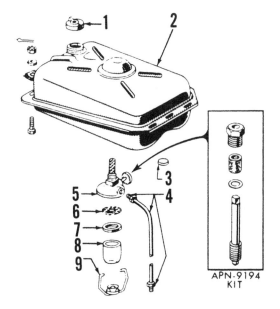

Fig. 161—Fuel tank and related parts.

1. Cap 9N9030
2. Tank 9N9002
3. Isolator 9N9121A
4. Tube 9N9282A
5. Bulb assy. 2N9155B
6. Screen 9N9161
7. Gasket 9N9160
8. Bowl 9N9162
9. Retainer 9N9166

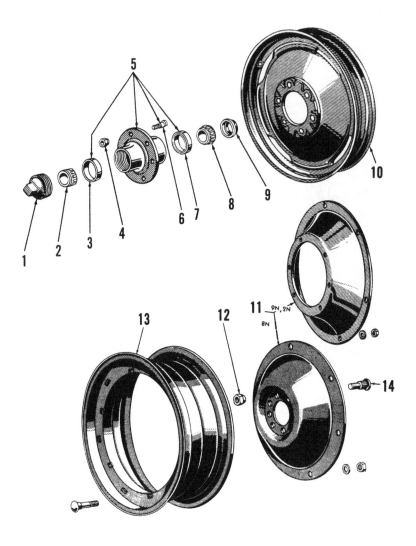

Fig. 162—Front and rear wheels and related parts.

1. Cap 2N1139B
2. Cone B1216
3. Cup B1217
4. Nut 2N1012
5. Hub assy. 9N1104A3
   (9N, 2N) 8N1104 (8N)
6. Bolt 781107
7. Cup B1202
8. Cone B1201
9. Seal 9N1190B
10. Wheel 9N1015A (9N, 2N)
    8N1015A (8N)
11. Rear disc 9N1122D or E
    (9N, 2N) 8N1122 (8N)
12. Nut 8N1134
13. Rim 9N1108D
14. Bolt 9N1117 (9N, 2N)
    8N1117 (8N)

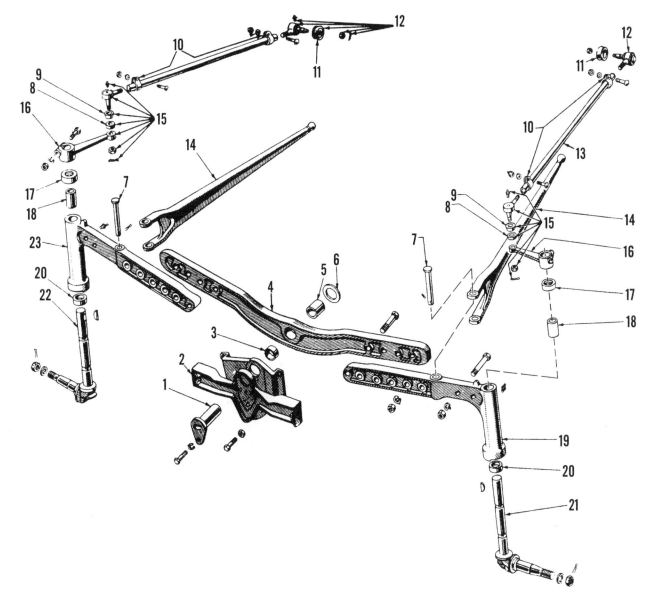

Fig. 163–Front axle and related parts.

1. Pin 2N3126D
2. Support 2m3029B
3. Spacer 8N3024 (8N)
4. Axle 9N3010A (9N, 2N) 8N3010A (8N)
5. Bushing 9N3039 (9N, 2N) 2N3039 (8N)
6. Spacer 8N3027 (8N)
7. Pin 357623S8
8. Seal 78 3332A (9N, 2N) 8N332 (8N)
9. Retainer 78 3336 (9N, 2N)
10. Clamp 8N3287
11. Cover 9N3324
12. End NAA33271A
13. Drag link 8N3314D

14. Radius rod 2N3405B (9N, 2N) 8N3405C (8N)
15. End 8N3270C
16. Arm 2N3130 (9N, 2N) 8N3130A (8N)
17. Seal 9N3528 (9N, 2N) C5NN3125A (8N)
18. Bushing 2N3109
19. Axle LH 9N3007A (9N, 2N) 8N3007A
20. Bearing 9N3123B
21. Spindle LH 2N3106 (9N, 2N) 8N3106B (8N)
22. Spindle RH 2N3105 (9N, 2N) 8N3105B (8N)
23. Axle RH 9N3006A (9N, 2N) 8N3006A (8N)

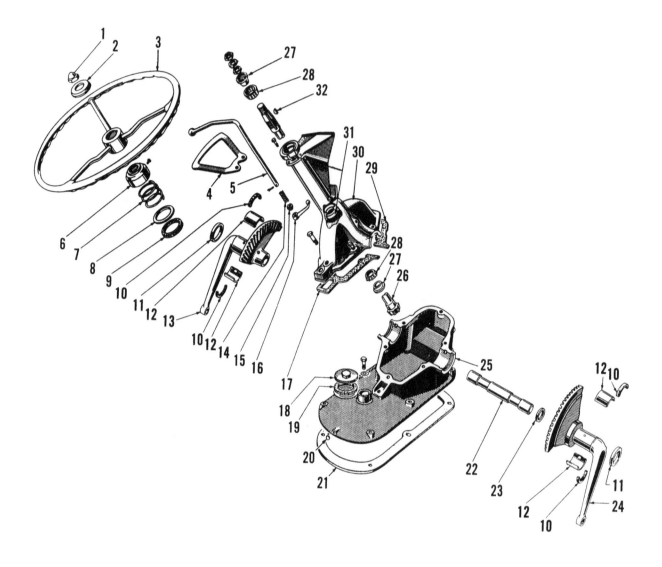

Fig. 164—Steering gear assembly for 9N and 2N models.

1. Nut 351103S
2. Washer 2N3673
3. Steering wheel 2N3600
4. Quadrant 9N9820
5. Throttle rod 9N9805
6. Cap 9N3669
7. Spring 9N3699
8. Retainer 9N3661
9. Seal 9N3660
10. Packing 9N3670
11. Washer 9N3543A, B or C
12. Bearing 9N3517
13. Sector 9N3527
14. Spring 2N9806
15. Retainer 9N9812
16. Lever 9N9807
17. Gasket 9N3593B

18. Plug 9N7010
19. Gasket 8N7011
20. Plug 74106S
21. Gasket 9N7223
22. Shaft 9N3540
23. Washer 9N3563A, B or C
24. Sector 9N3526
25. Housing 9N3550C
26. Pinion 9N3525
27. Cone 9N3573
28. Bearing 68 3571A
29. Gasket 9N3594
30. Housing 9N3551D
    (foot starter button),
    9N3551E (inst. panel button)
31. Cup B3552
32. Key B3609

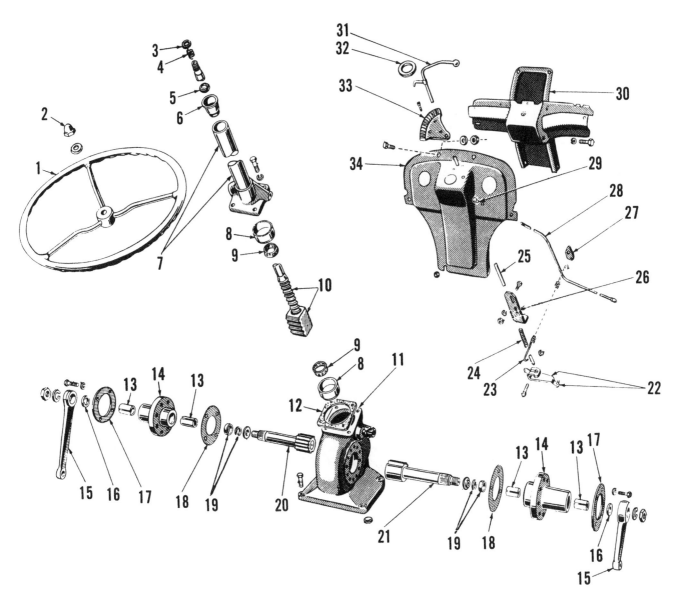

*Fig. B165—Steering gear and instrument panel
for 8N tractors before serial number 216989.*

1. Wheel 8N3600
2. Nut 35114S
3. Seal 8N3570
4. Spring 8N3520
5. Seat 8N3518
6. Bearing 8N3517
7. Cover 8N3545A
8. Cup 8N3552
9. Bearing 8N3571
10. Shaft 8N3575A
11. Housing 8N3548A
12. Shim 8N3595A, B, C or D
13. Bushing 8N3576A
14. Side cover 8N3561A
15. Arm 8N3590A
16. Seal 8N3586
17. Gasket 8N3579
18. Gasket 8N3581A
19. Seal D5UZ3591A
20. Sector LH 8N3527A
21. Sector RH 8N3526A
22. Lever 8N9807
23. Spring 8N9827
24. Spring 8N9806
25. Rod 8N9805A4
26. Bracket 8N9796
27. Plate 8N9855
28. Choke rod 8N9700
29. Knob 9N9703A
30. Support 8N3532B
31. Lever 8N9805A4
32. Grommet 8N3578
33. Quadrant 8N9889A2
34. Panel 8N3512C

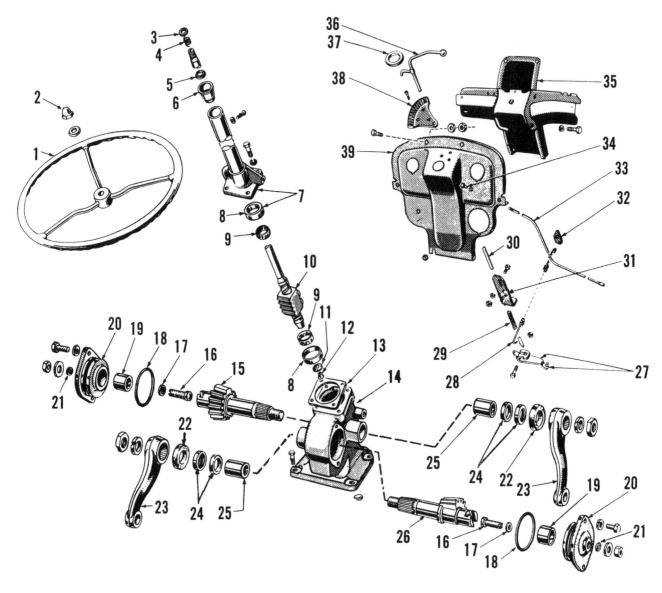

**Fig. 166—Steering gear and instrument panel
for 8N tractors after serial number 216989.**

1. Wheel 8N3600
2. Nut 35114S
3. Seal 8N3570
4. Spring 8N3520
5. Seat 8N3518
6. Bearing 8N3517
7. Cover 8N3545B
8. Cup 8N3552
9. Bearing 8N3571
10. Shaft 8N3575B
11. Retainer 8N33581
12. Eyelet 356937S
13. Shim 8N3595A, B, C or D
14. Housing 8N3548B
15. Sector LH 8N3527B
16. Adjuster 8N33545
17. Shim 8N33544A, B, C or D
18. Seal 8N3581B
19. Bushing 8N3576B
20. Side cover 8N3561B

21. Seal 8N33546
22. Seal 8N3586
23. Arm NAAA3590A
24. Seal D5UZ3591A
25. Bushing 8N3553
26. Sector RH 8N3526B
27. Lever 8N9807
28. Spring 8N9827
29. Spring 8N9806
30. Rod 8N9805A4
31. Bracket 8N9796
32. Plate 8N9855
33. Choke rod 8N9700
34. Knob 9N9703A
35. Support 8N3532B
36. Lever 8N9805A4
37. Grommet 8N3578
38. Quadrant 8N9889A2
39. Panel 8N3512C

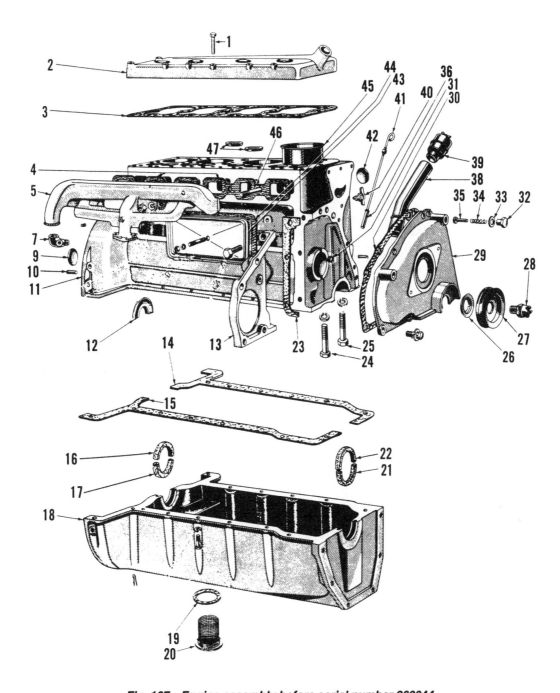

Fig. 167—Engine assembly before serial number 263844.

A. Overhaul gasket
kit 8N6008
B. Oil pan gasket
kit 9N6781
1. Bolt 8N6065A
2. Head 8N6050A
3. Gasket 8N6051
4. Gasket 9N9448
5. Manifold 9N9425
7. Elbow 8N6722
9. Plug 91A6266
10. Dowel B6397A
11. Block 8N6010C
12. Seal 8N6335A
13. Cover 9N6017
14. Gasket 8N6711

15. Gasket 8N6710
16. Seal 8N6701
17. Seal 8N6702
18. Oil pan 9N6675
19. Gasket 8N6734
20. Screen 9N6730
21. Seal 8N6707
22. Seal 8N6700
23. Gasket 8N6018
24. Bolt 8N6346A
25. Bolt 8N6345A
26. Slinger 40 6310A2
27. Pulley 2N6312
28. Ratchet 2N6319A1
29. Cover 9N6019B
30. Gasket 9N6020A

31. Dowel 18 6397
32. Nut 68 6666
33. Gasket 1GA6669
34. Spring 9N6654
35. Plunger 01A6663
36. Tube 9N7020
38. Pipe 8N6763B
39. Cap 8N6766B
40. Valve 9N8115B
41. Dipstick 9N6750A
42. Plug 353454S
43. Cover 7HA6520
44. Gasket 7HA6521
45. Sleeve 8N6055B
46. Stud 88403S
47. Insert 40 6057B int.
52 6057B exh.

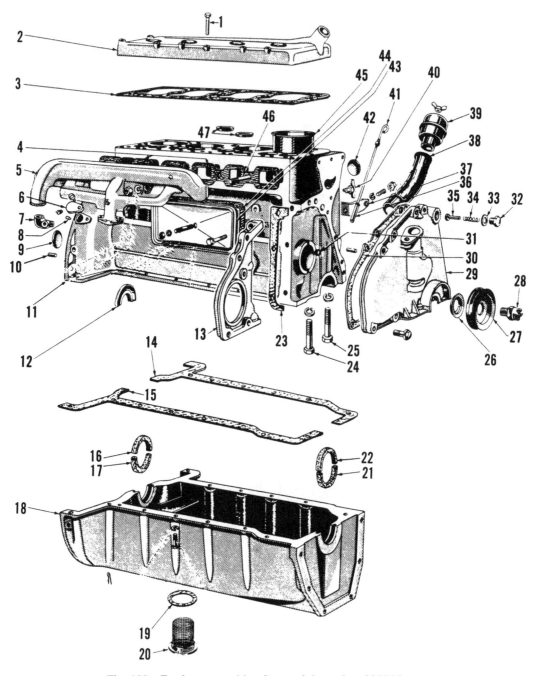

Fig. 168—Engine assembly after serial number 263844.

A. Overhaul gasket
   kit 8N6008
B. Oil pan gasket
   kit 9N6781
1. Bolt 8N6065A
2. Head 8N6050A
3. Gasket 8N6051
4. Gasket 9N9448
5. Manifold 9N9425
7. Elbow 8N6722
9. Plug 91A6266
10. Dowel 18 6397
11. Block 8N6010C
12. Seal 8N6335A
13. Cover 8N6017B
14. Gasket 8N6711
15. Gasket 8N6710

16. Seal 8N6701
17. Seal 8N6702
18. Oil pan 9N6675
19. Gasket 8N6734
20. Screen 9N6730
21. Seal 8N6707
22. Seal 8N6700
23. Gasket 8N6018
24. Bolt 8N6346A
25. Bolt 8N6345A
26. Slinger 40 6310A2
27. Pulley 2N6312
28. Ratchet 2N6319A1
29. Cover 8N6019B
30. Gasket 8N6020C
31. Dowel 18 6397
32. Nut 68 6666

33. Gasket 1GA6669
34. Spring 9N6654
35. Plunger 01A6663
36. Tube 9N7020
37. Bracket 8N6129
38. Pipe 8N6763B
39. Cap 8N6766B
40. Valve 9N8115B
41. Dipstick 9N6750A
42. Plug 353454S
43. Cover 7HA6520
44. Gasket 7HA6521
45. Sleeve 8N6055B
46. Stud 88403S
47. Insert 40 6057B int.
    52 6057B exh.

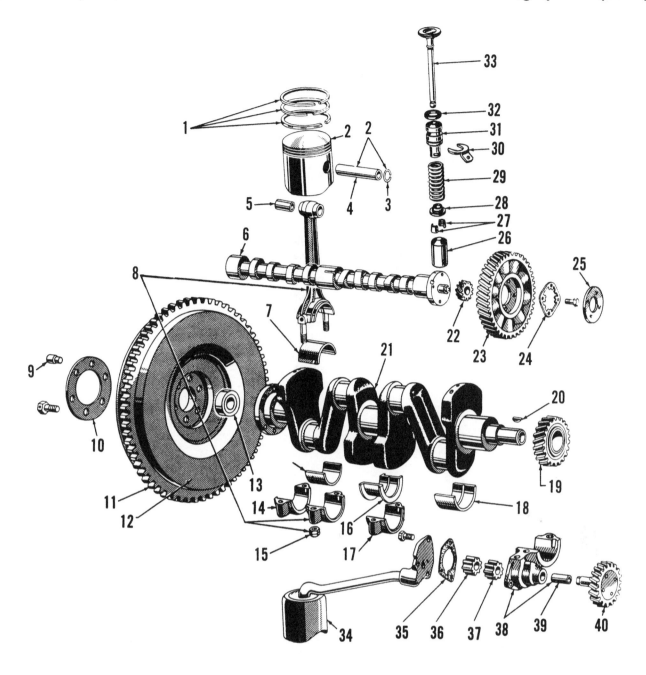

Fig. 169—Exploded view of engine. (BSN—before serial number; ASN—after serial number.)

1. Rings EBPN6149A
   (steel piston) 8N6149B
   (alum. piston)
2. Piston 8N6108B
3. Retainer 78 6140
4. Pin 19A 6135
5. Bushing 21A 6207B3
6. Camshaft 2NC6250B
   (BSN 263843) 8N6250
   (ASN 263843)
7. Bearing 9N6211A
8. Rod 9N6200B
9. Dowel B6387A
10. Retainer 7RA7609
11. Ring gear 9N6384
12. Flywheel 8N6375
13. Bearing C5NN7600A
14. Cap 9N6325

15. Nut C9AZ6212B
16. Bearing 9N6331A
17. Cap 9N6330
18. Bearing 9N6333A
19. Gear 48 6306A
20. Key 1/4 x 1-1/8
21. Crankshaft 9N6303A
22. Gear 7RA6255
23. Gear 48 6256B
    (BSN263843) 7RA6256A
    (ASN 263843)
24. Ring 91A6258
25. Spring 8N6276
    (BSN 263843)
26. Tappet AONN6500A
27. Lock 7HA6518
28. Retainer 8RT6514B
    exh. 8BA6514 int.

29. Spring 8N6513
30. Retainer 406512
31. Guide 8BA6510B
32. Seal 8BA6571
33. Valve 8N6505A exh.
    8BA6507A int.
34. Screen 8N6615
35. Gasket 9N6619
36. Gear 52 6610B (BSN 247571)
    8N6610 (ASN 247571)
37. Gear 52 6614B (BSN 247571)
    8N6614 (ASN 247571)
38. Housing 9N6603 (BSN 247571)
    8N6603 (ASN 247571)
39. Bushing 9N6612A
40. Gear 9N6608A (BSN 247571)
    8N6608A (ASN 247571)

113

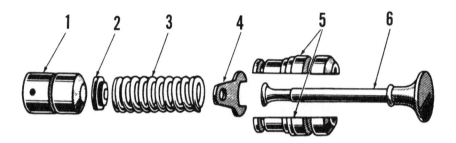

Fig. 170—Engine valve components for tractors before serial number 42162. These parts can be replaced by late style valve and tappet kit 8N6546B.

1. Tappet 8BN6500A
2. Retainer 40 6514B
3. Spring 2NC6513
4. Retainer 40 6512
5. Guide 40 6510C
6. Valve 40 6505A int.
   9N6505A exh.

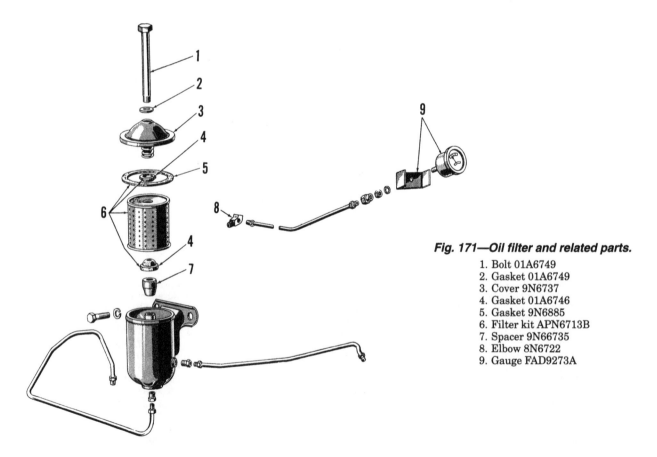

Fig. 171—Oil filter and related parts.

1. Bolt 01A6749
2. Gasket 01A6749
3. Cover 9N6737
4. Gasket 01A6746
5. Gasket 9N6885
6. Filter kit APN6713B
7. Spacer 9N66735
8. Elbow 8N6722
9. Gauge FAD9273A

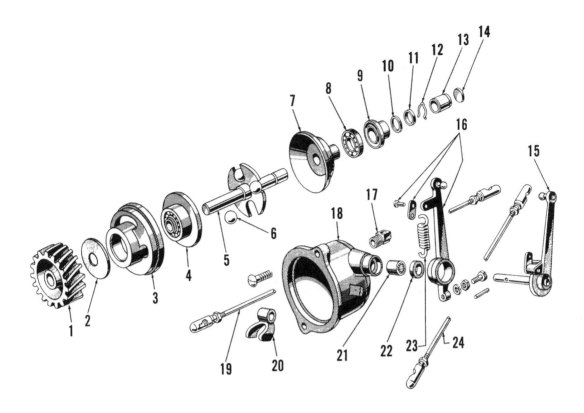

Fig. 172—Governor and related parts.

1. Gear 8N18222
2. Washer 357319S
3. Base 8N18191B
4. Race 9N18186B
5. Shaft 9N18187A (1939-1950)
   8N18187A (1950-1952)
6. Ball 358081S
7. Race 9N18188B
8. Bearing 9N18192
9. Base 9N18194
10. Shim 9N18241
11. Washer 356430S
12. Clip 9N18195
13. Bushing 9N18184
14. Washer 356491S
15. Lever 9N18189
16. Lever 9N18190
    (9N, 2N) 8N18190 (8N)
17. Elbow 87381S
18. Housing 8N18185B
19. Rod 9N9815B (9N, 2N)
    8N9815 (8N)
20. Fork 9N18235
21. Bearing 9N18182
22. Seal 9N18183
23. Spring 9N18196
24. Rod 9N9818A

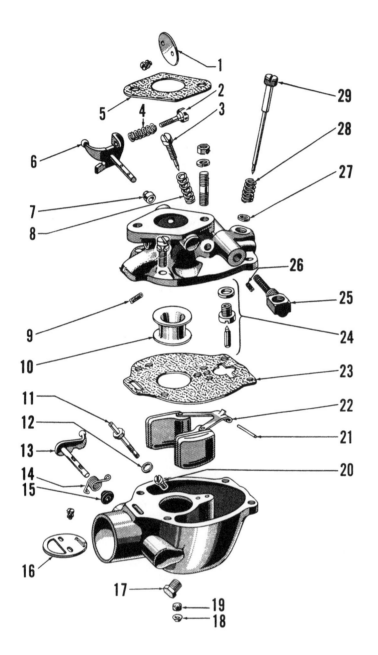

Fig. 173—Carburetor assembly. Carburetor repair kit part number is APN9590A.

1. Throttle 8N9585
2. Screw 34093S
3. Needle 9N9541
4. Spring 9N9589
5. Gasket 9N9447
6. Shaft 9N9581A
7. Packing 8N9622
8. Spring 9N9578
9. Idle jet 9N9596
10. Venturi C5NN9586B
11. Nozzle 9N9530
12. Gasket 9N9608
13. Shaft 9N9546
14. Spring 9N9539
15. Packing 8N9689
16. Choke 9N9549
17. Plug 87650S36
18. Plug 9N9692
19. Strainer 9N9694A
20. Main jet 9N9533
21. Pin 9N9558A
22. Float 9N9550A
23. Gasket C3NN9516A
24. Valve C3NN9564A
25. Elbow 9N9553
26. Jet 9N9914 (9N, 2N)
    8N9914 (8N)
27. Gasket 9N9563
28. Spring 9N9540C
29. Needle 9N9565C

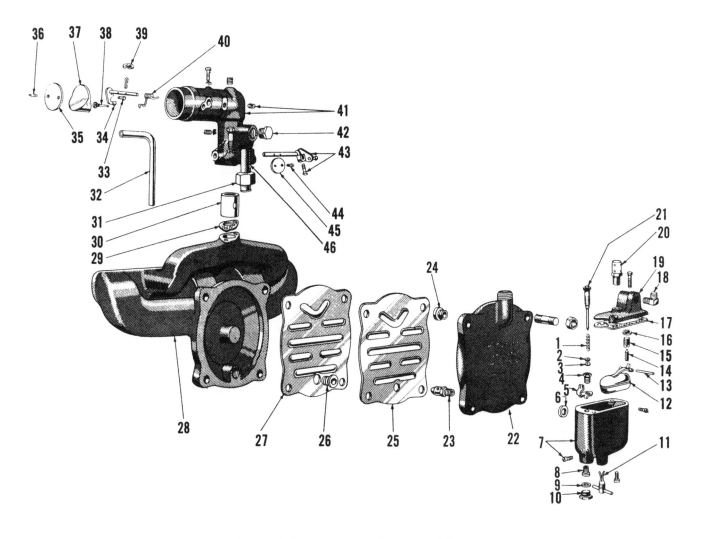

Fig. 174—Kerosene vaporizer assembly.

1. Spring 9NAN9540
2. Washer 9NAN9643
3. Packing 9NAN9642
4. Bushing 9NAN9644
5. Ratchet 9NAN9536
6. Gasket 9NAN9956
7. Chamber 9NAN9951
8. Bushing 9NAN9604
9. Gasket 9NAN9563
10. Plug 9NAN9522
11. Valve 9NAN9523
12. Float 9NAN9550
13. Pin 9NAN9558
14. Valve 78 9566
15. Seat 9NAN9567
16. Gasket 9NAN9569
17. Gasket 9NAN9519
18. Elbow 9NAN9697
19. Cover 9NAN9953
20. Vent 9NAN9954
21. Needle 9NAN9565
22. Cover 9NAN9968
23. Jet 9NAN9971

24. Bushing 9NAN9970
25. Plate 9NAN9969
26. Bushing 9NAN67
27. Plate 9NAN9966
28. Manifold 9NAN9428
29. Gasket 9NAN9516
30. Venturi 9NAN9586
31. Nut 9NAN9959
32. Tube 9NAN9960
33. Screw 9NAN9963B
34. Shaft 2NAN9546
35. Plate 9NAN99545
36. Pin 9NAN9972
37. Valve 9NAN9961
38. Shaft 9NAN9962B
39. Plate 9NAN9964
40. Spring 9NAN9624
41. Plug 9NAN9521
42. Plug 9NAN9522
43. Shaft 9NAN9581
44. Screw 40 9586
45. Throttle 9NAN9585
46. Tube 9NAN9958

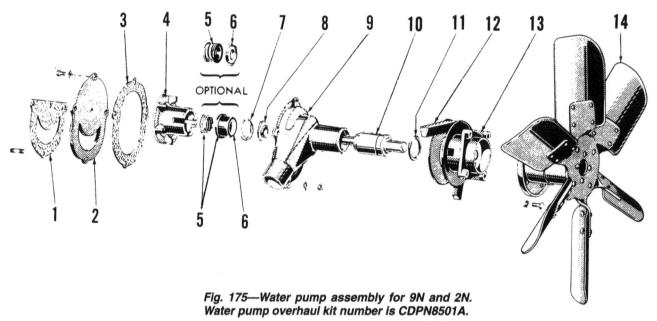

Fig. 175—Water pump assembly for 9N and 2N.
Water pump overhaul kit number is CDPN8501A.

1. Gasket 9N8513
2. Cover 8N8508
3. Gasket 9N8507
4. Impeller 78 8512
5. Seal 68 8551A
6. Washer 9N8557
7. Ring 68 8574
8. Bushing 9N8520
9. Housing 2N8505
10. Bearing 78 8530
11. Ring OBA8576A
12. Belt 2N8620B
13. Pulley 9N8606
14. Fan 9N8600D

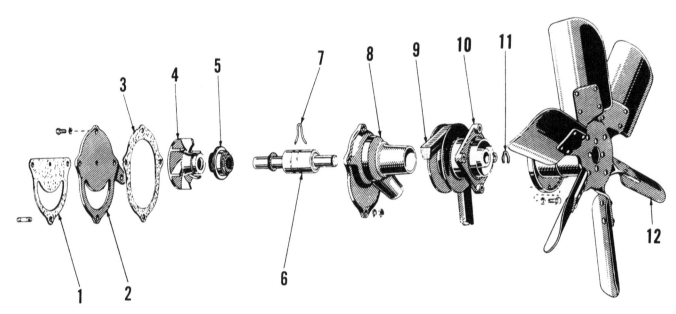

Fig. 176—Water pump assembly for 8N.
Water pump overhaul kit number is CDPN8501A.

1. Gasket 9N8513
2. Cover 8N8508
3. Gasket 9N8507
4. Impeller EAF8516A
5. Seal D5TE8564B1A
6. Bearing 2N8578
7. Retainer 7HA8576
8. Housing 2N8505
9. Belt 8N8620
10. Pulley 2N8610
11. Ring 2GA8531
12. Fan 9N8600D

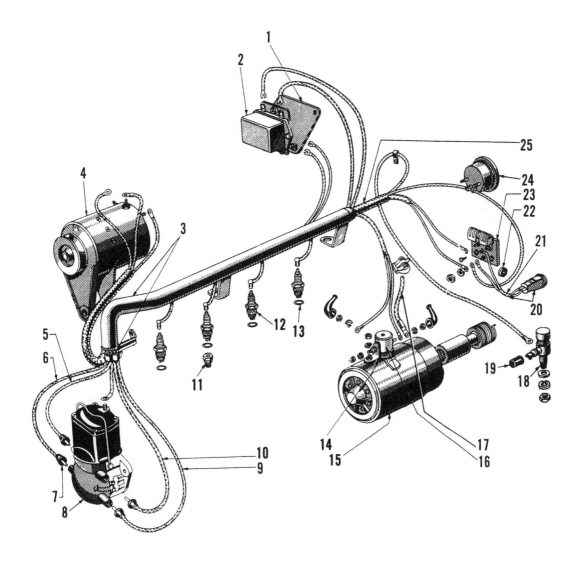

Fig. 177—Electrical wiring and related parts for 9N, 2N and 8N before serial number 263844.

1. Bracket 2N10599
2. Cut-out BONN10505A
3. Conduit 2N12112
4. Generator 8N10000BLPR
5. Wire 9N12287B
6. Wire 2N12284
7. Insulator FDS12113A
8. Distributor 9N12100
9. Wire 9N12283
10. Wire 9N12286
11. Adapter 2N12426
12. Spark plug AL7C
13. Gasket 01A12410A
14. Bar 8N11518A
15. Starter 8N11001R
16. Relay 8N11450
17. Cable 9N14300C
18. Switch 8N11500
19. Cover 8N11113
20. Ignition switch 8N3679C
21. Wire 9N14498
22. Grommet 11A14605A
23. Resistor A8NN12250A
24. Ammeter 9N10850B
25. Harness 8N14401B

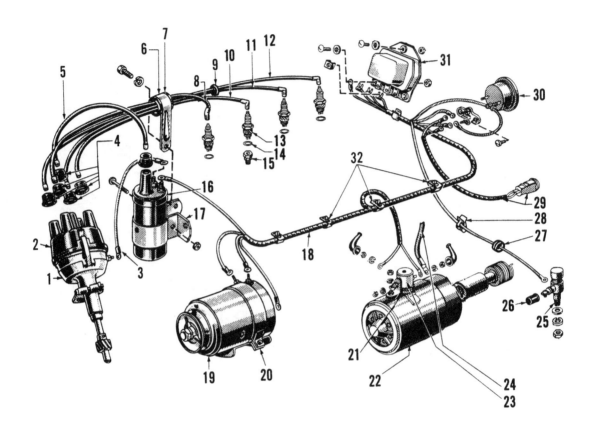

***Fig. 178—Electrical wiring and related
parts for 8N after serial number 263843.***

1. Distributor 8N12127B
2. Cap NCA12106A
3. Wire FDN14302A
4. Insulator FDS12113A
5. Wire 8N12298
6. Bracket 8N12112B
7. Grommet 7RA12297
8. Wire 8N12287C
9. Retainer 7RA12271
10. Wire 9N12284
11. Wire 8N12283B
12. Wire 8N12286B
13. Spark plug AL7C
14. Gasket 01A12410A
15. Adapter 2N12246
16. Coil E8TF12029BA
17. Bracket 8N12043
18. Harness 8N14401C
19. Generator 8N10001LPR
20. Bracket 8N10151C
21. Bar 8N11518B
22. Starter 8N11001R
23. Relay 8N11450
24. Cable 9N14300C
25. Starter switch 8N11500
26. Cover 8N11113
27. Grommet 11A14605A
28. Clip 356668S32
29. Key switch 8N3679C
30. Ammeter 9N10850B
31. Regulator 8N10505C
32. Clips 7RC14197

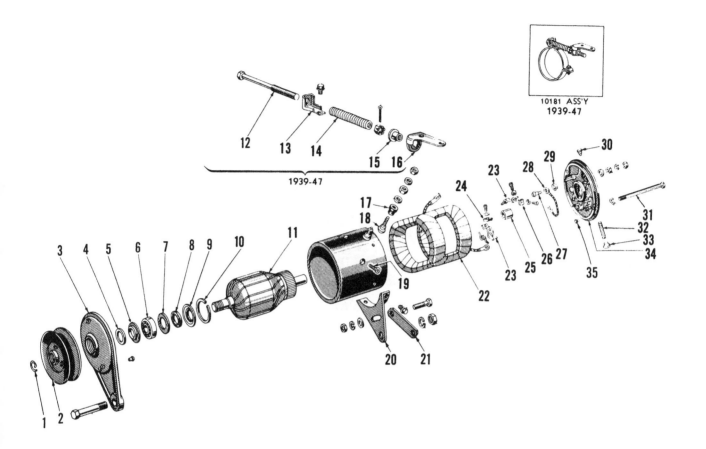

Fig. 179—Exploded view of 3-brush generator. Brush set part number is 40-10043.

1. Ring 9N10096C
2. Pulley 9N10130B
3. End plate assy. 9N10138C
4. Seal 9N10122
5. Retainer 9N10121
6. Bearing E27N10094B
7. Retainer 9N10214
8. Seal 9N10212
9. Retainer 9N10098
10. Ring 9N10163
11. Armature 8N10005A
12. Bolt 2N10176A
13. Bracket 2N10167A
14. Spring 2N10177
15. Nut 2N10180
16. Bracket 2N10178
17. Insulator 81A10206
18. Screw 11A10211B, insulator 81A10208

19. Screw 68 10044
20. Bracket 8N10151A
21. Arm 8N10145
22. Coil 8N10175
23. Holder 40 10051
24. Brush 40 10069
25. Bushing 9N10128B
26. Brush 40 10070
27. Screw11A10211B
28. Wire 9N10036
29. Insulator 81A10208
30. Oil cup B10141A
31. Bolt C2AF10120A
32. Wick 18 10147
33. Plug 18 10146
34. End plate 9N10129B
35. Dowel 18 10088

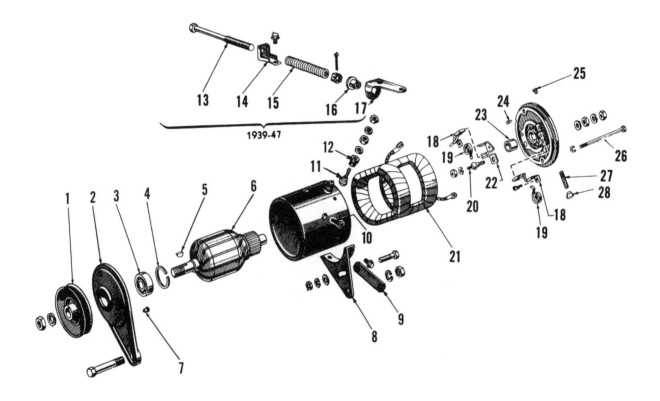

*Fig. 180—Exploded view of 2-brush
generator prior to serial number 263844.*

1. Pulley 9N10130B
2. Plate 8N10138A
3. Bearing E27N10094B
4. Ring 9N10163
5. Key 5/32 x 5/8
6. Armature 9N10005C
7. Dowel 18 10088
8. Bracket 8N10151A
9. Arm 8N10145
10. Screw 68 10044
11. Screw 11A10211B,
    insulator 81A10208
12. Insulator 81A10206
13. Bolt 2N10176A
14. Bracket 2N10167
15. Spring 2N10177
16. Nut 2N10180
17. Bracket 2N10178
18. Brush
19. Spring
20. Screw 11A10211B
21. Coil 9N10175C
22. Holder 40 10051
23. Bushing 9N10128B
24. Dowel 18 10088
25. Oil cup B10141A
26. Bolt 18 10120A
27. Wick 18 10147
28. Plug 18 10151A

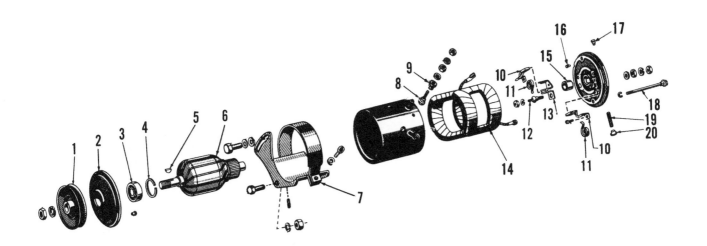

Fig. 181—Exploded view of 2-brush
generator after serial number 263843.

1. Pulley 8N10130A
2. Plate 8N10138B
3. Bearing 65391
4. Ring 01A10163
5. Key 5/32 x 5/8
6. Armature FAA10005BR
7. Bracket 8N10151C
8. Screw 11A10211B,
   insulator 81A10208
9. Insulator 81A10206
10. Brush FAA10069A

11. Spring 91A10057
12. Screw 11A10211B
13. Holder 40 10051
14. Coil 8N10175B
15. Bushing 78 10128
16. Dowel 18 10088
17. Oil cup B10141A
18. Bolt C2AF10120A
19. Wick 18 10147
20. Plug 18 10146

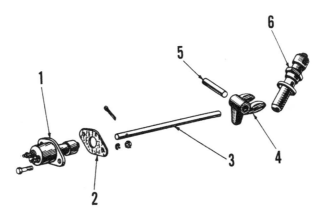

Fig. 182—Starter switch and button for 9N and 2N.
(BSN—before serial number; ASN—after serial number.)

1. Switch 9N11450B (BSN 12500),
   9N11467 (ASN 12499)
2. Gasket 9N11506
3. Rod 9N11512 (ASN 12499)
4. Lever 2N11515 (ASN 12499)
5. Pin 9N11511 (ASN 12499)
6. Button 9N11506 (ASN 12499)

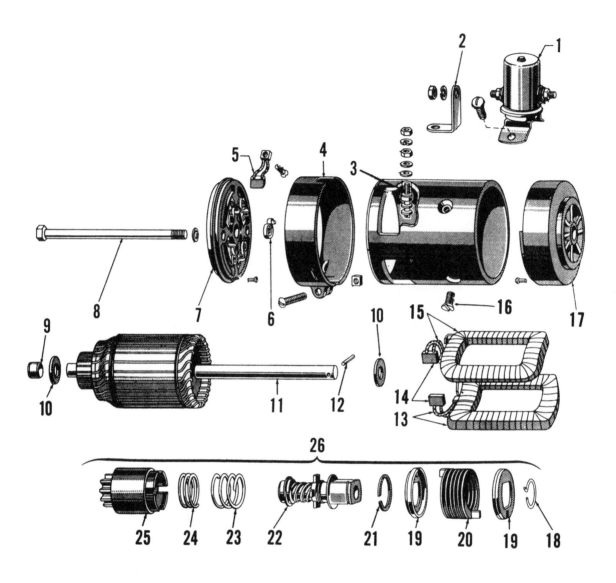

Fig. 183—Starter motor and related parts.

1. Relay 8N11450
2. 8N11518
3. Terminal C7NF11102A,
   insulator 18 11107B
4. Cover 18 10088
5. Brush 18 11056
6. Spring D1AF11059AA
7. Plate C3AF11050A
8. Bolt 18 11091
9. Bearing C4TF11052A
10. Washer 18 11036
11. Armature D1AZ11059A
12. Pin 52 11365B
13. Coil FAC11083A
14. Brushes 18 11055
15. Coil 18 11085
16. Screw 68 1044
17. Plate 8N11130
18. Ring 52 11373
19. Plate 52 11372B
20. Spring 52 11375
21. Ring 52 1170
22. Drive shaft 52 11366B
23. Spring 52 11369
24. Spring 52 11368
25. Pinion 52 11367
26. Drive assy.
    D8NN11350BA

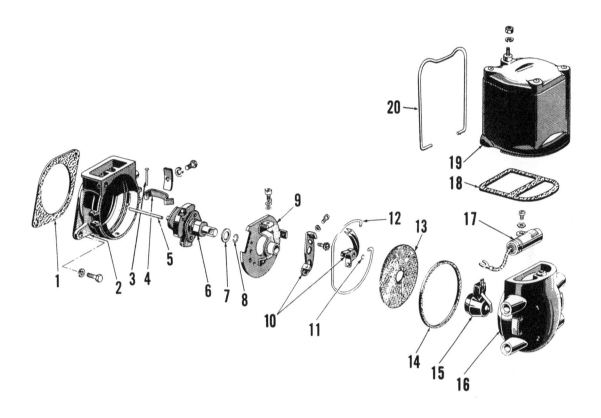

Fig. 184—Distributor and coil assembly before serial number 263844.

1. Gasket 9N12143
2. Housing 9N12139
3. Rivet B12145
4. Clamp 9N12144
5. Wick 9N12133
6. Cam 9N12187
7. Washer 91A12179
8. Ring 91A12177
9. Breaker plate 9N12151
10. Points A0NN12107A
11. Ring 91A12212
12. Spring 91A12146
13. Cover 9N12276B
14. Gasket 91A12114
15. Rotor B2NN12200A
16. Cap 9N12106C
17. Condenser A0NN12300A
18. Gasket 9N12140A
19. Coil 9N12024
20. Bail 9N12137

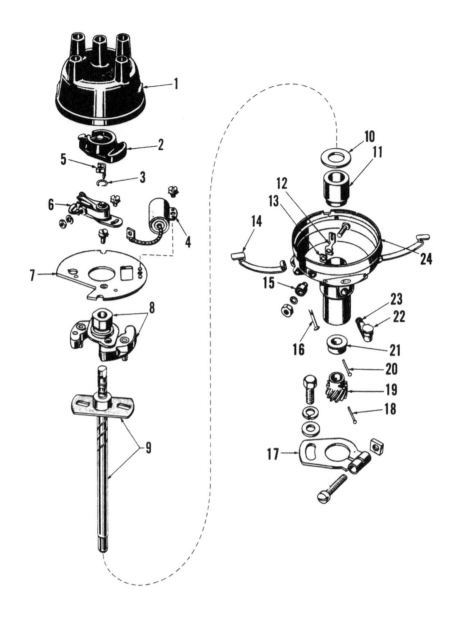

Fig. 185—Distributor assembly after serial number 263843.
Points and condenser kit part number is CPN12000A.

1. Cap NCA12106A	13. Insulator 7RA12234
2. Rotor C0NN12200A	14. Clamp 7RA12144
3. Ring 97468S	15. Bushing 7RA12233
4. Condenser FET12300A	16. Rivet B12145
5. Spring 8N12213	17. Arm 7HA12273
6. Points D8NN12171AA	18. Rivet 1/8 x 11/16
7. Plate FAD12152A	19. Gear 7RA12390C
8. Cam 8N12176	20. Rivet 1/8 x 13/16
9. Shaft 8N12175	21. Collar 7HA12195
10. Washer B4A12179A	22. Oiler 8EQ12135A
11. Bushing 8N12120	23. Wick 40 12141
12. Conductor 8N12209	24. Base 8N12130C

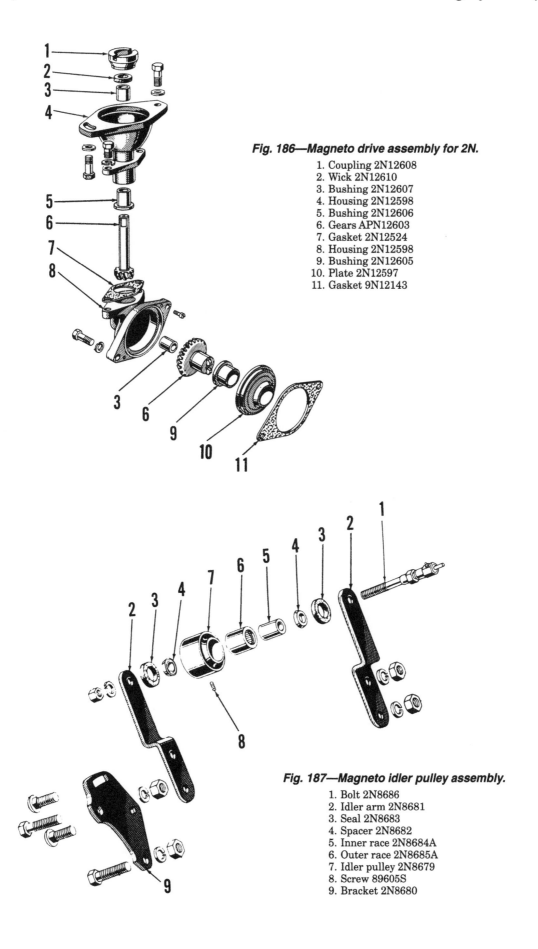

Fig. 186—Magneto drive assembly for 2N.

1. Coupling 2N12608
2. Wick 2N12610
3. Bushing 2N12607
4. Housing 2N12598
5. Bushing 2N12606
6. Gears APN12603
7. Gasket 2N12524
8. Housing 2N12598
9. Bushing 2N12605
10. Plate 2N12597
11. Gasket 9N12143

Fig. 187—Magneto idler pulley assembly.

1. Bolt 2N8686
2. Idler arm 2N8681
3. Seal 2N8683
4. Spacer 2N8682
5. Inner race 2N8684A
6. Outer race 2N8685A
7. Idler pulley 2N8679
8. Screw 89605S
9. Bracket 2N8680

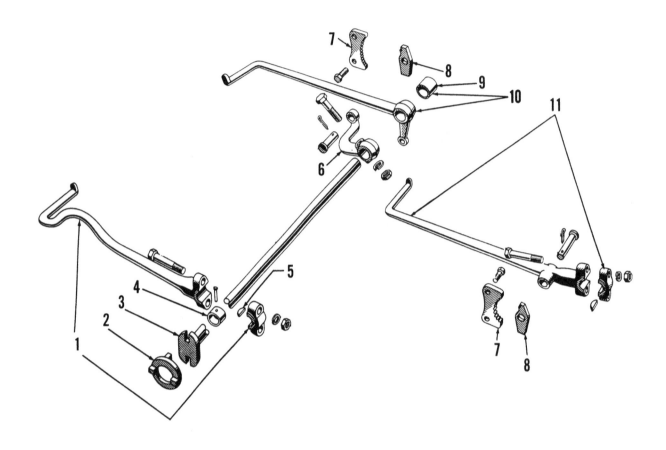

Fig. 188—Clutch and brake controls for 9N and 2N.

1. Brake pedal RH 9N2455A
2. Wedge 9N2050
3. Camshaft 9N2247
4. Bushing 9N2248
5. Key 1/4 x 1-1/8
6. Arm 9N2324
7. Sector 9N2796
8. Pawl 9N2786
9. Bushing 9N7508
10. Clutch pedal 9N7520
11. Brake pedal LH 9N2456A

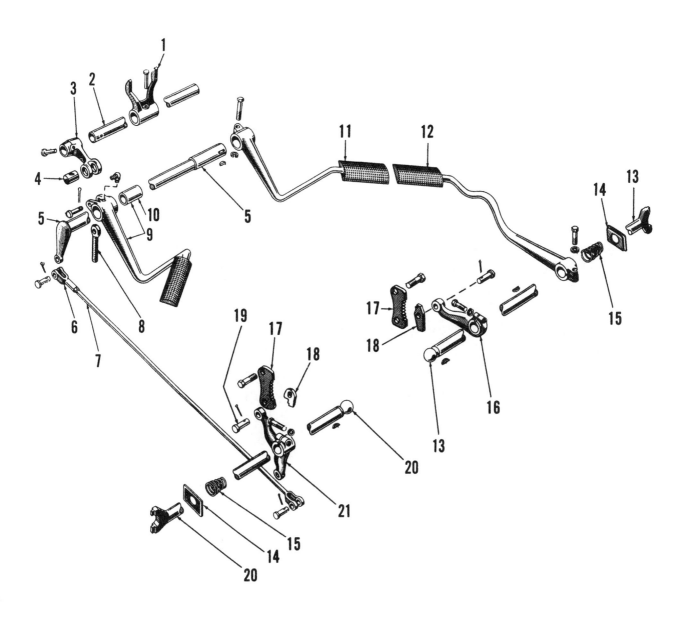

Fig. 189—Clutch and brake controls for 8N.

1. Clutch fork 48 7515
2. Shaft B7510A2
3. Arm 9N7511
4. Pin 8N7533
5. Shaft NCA2458A
6. Clevis 3/8 x 2-1/2
7. Rod 8N2465
8. Eyebolt 355584S8
9. Clutch pedal NAA7519B
10. Bushing 8N7508B
11. Brake pedal LH 8N2456A
12. Brake pedal RH 8N2455
13. Camshaft 8N2246
14. Cover 8N2224
15. Spring 8N2322
16. Arm 8N2324
17. Sector 8N2796
18. Pawl 9N2786
19. Pin 1/2 x 1-3/8
20. Camshaft 8N2247
21. Lever 8N2315

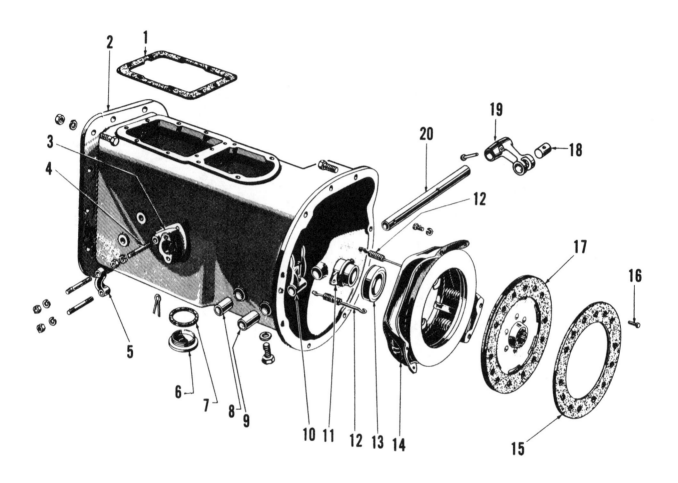

Fig. 190—Transmission case, clutch and related parts.

1. Gasket E6NN7223BA
2. Case 9N7005B, 8N7005C
3. Spacer C5NN3N442C
4. Stud 88398S
5. Cap C7NN3N406A
6. Drain plug 8N7010
7. Gasket 8N7011
8. Bushing C0NN2477A
9. Bushing 40 7508A
10. Fork 48 7515

11. Hub 9N7561
12. Spring 9N7562
13. Bearing C0NN7580A
14. Pressure plate 8N7563
15. Facing 40 7549
16. Rivet 9/64 x 7/32
17. Clutch disc 91A7550
18. Pin 8N7533
19. Arm 8N7511B
20. Shaft B7510A2

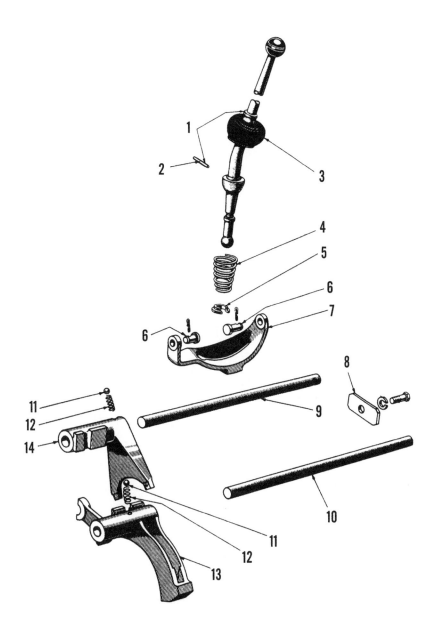

Fig. 191—Gearshift and related parts for 9N and 2N.

1. Lever 9N7209
2. Pin 9N7221B
3. Seal C5NN7277C
4. Spring 9N7227B
5. Seat 9N7228A2
6. Pin 3/16 x 1
7. Interlock 9N7229B
8. Plate 9N7225
9. Rob 9N7240A
10. Rod 9N7241
11. Ball 353075S
12. Spring 79 7234
13. Fork 9N7230A (BSN 12500),
    9N7230B (ASN 124992)
14. Fork 9N7231B

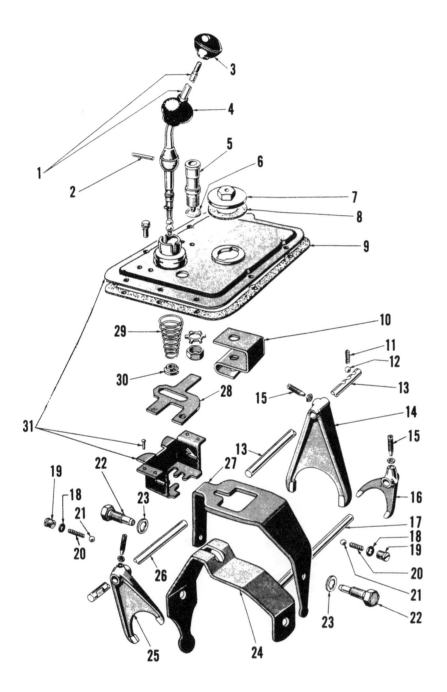

Fig. 192—Gearshift and related parts for 8N.

1. Lever C5NN7202B
2. Pin 9N7221B
3. Knob C5NN7213A
4. Seal C5NN7277C
5. Switch 8N11500
6. Gasket 8N11529
7. Cap 8N7485
8. Gasket 8N7037
9. Gasket E6NN7223BA
10. Support 8N7257
11. Spring 9N7228A2
12. Ball 353075S
13. Rail D8NN7240AB
14. Fork NCA7230A
15. Screw 8N7245
16. Fork 8N7231
17. Rail 8N7242
18. Gasket E5NN7Z307AA
19. Plug 74 7238F
20. Spring 79 7234
21. Ball 353075S
22. Seat 8N7238
23. Seal D9NN7268AA
24. Plate 8N7225
25. Fork C5NN7N236B
26. Rail D3NN7241A
27. Plate 8N7216
28. Latch 8N7229
29. Spring 9N7227B
30. Seat 9N7228A2
31. Cover NAA7211A

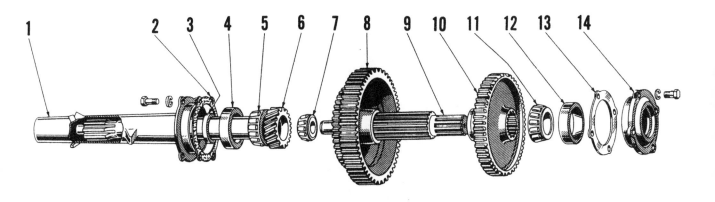

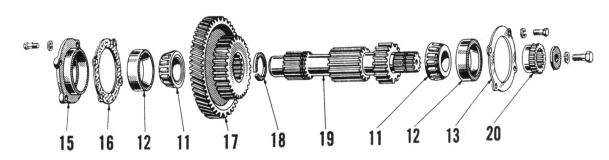

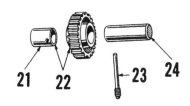

Fig. 193—3-speed transmission parts for 9N and 2N.

1. Retainer 9N7050	13. Shim C3NN7135
2. Gasket 9N7086	14. Retainer 9N7085
3. Seal 8N7052A	15. Retainer 9N7134
4. Cup 9N7067	16. Gasket 9N7086
5. Cone 9N7066	17. Gear 2N7113A
6. Input shaft 2N7017A	18. Ring 9N7070
7. Bearing 9N7120	19. Countershaft 9N7111
8. 1st & 3rd Gear 9N7100	20. Hub 9N7156
9. Main shaft 9N7061	21. Bushing 9N7143A
10. 2nd & rev. Gear 9N7101	22. Idler gear 9N7141
11. Cone 9N7066	23. Pin 9N7155
12. Cup 9N7067	24. Shaft 9N7140

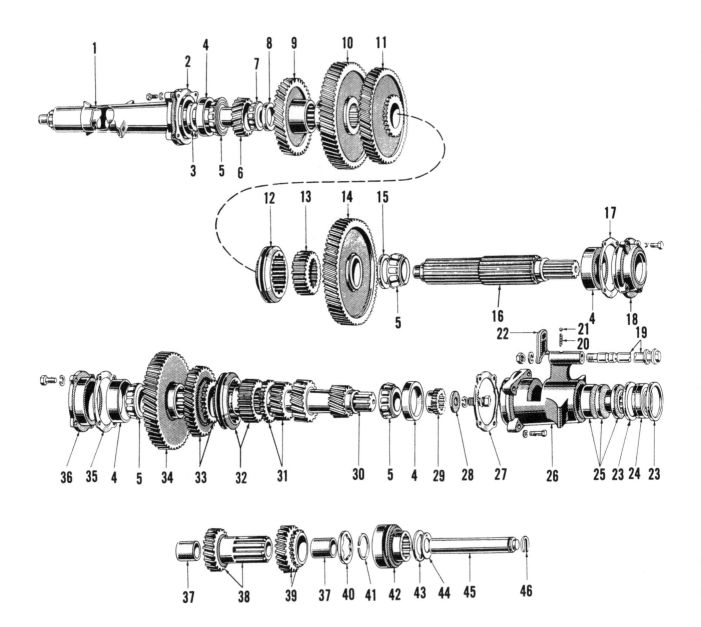

Fig. 194—4-speed transmission parts for 8N.

1. Retainer 9N7050	16. Main shaft 8N7061	31. Gear B9NN7114A
2. Gasket 9N7086	17. Shim C3NN7135	32. Connector 8N7105B
3. Seal 8N7052A	18. Retainer 9N7085	33. Gear 8N7112
4. Cup 9N7067	19. PTO rail 9N719	34. Gear 8N7113B
5. Cone 9N7066	20. Spring 79 7234	35. Gasket 9N7086
6. Input shaft 2N7017A	21. Ball 353075S	36. Retainer 9N7134
7. Bearing 9N7120	22. Stop NAA722A	37. Bushing 8N7143A
8. Washer 8N7071B	23. Ring 9N754	38. Gear 8N7141
9. 4th Gear 8N7110	24. Bearing 9N715C	39. Gear 8N7144
10. 2nd Gear B9NN7A381B	25. Sleeve 2N717	40. Thrust washer 8N7148
11. 3rd Gear 8N7102	26. PTO support 8N770	41. Ring 9N7070
12. Coupling 8N7106	27. Shim C3NN7135	42. Coupling 8N7145
13. Connector 8N7108	28. Washer 351402S	43. Thrust washer 8N7149
14. 1st Gear 8N7100	29. Hub 9N716	44. Thrust washer 8N7029
15. Washer 8N7071A	30. Countershaft 8N7111	45. Shaft NAA7140A
		46. Ring 8N7155

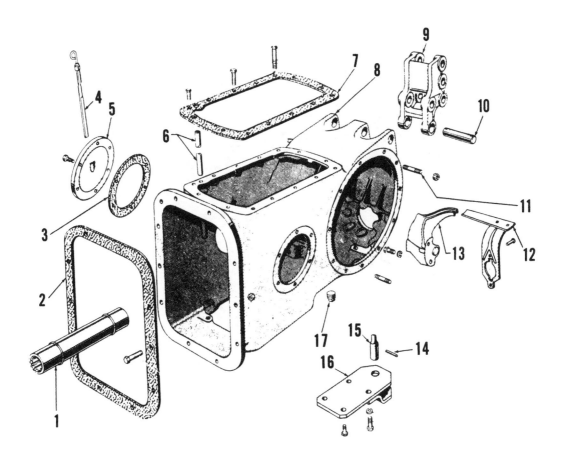

Fig. 195—Center housing and related parts.

1. Coupling 8N4602B
2. Gasket 9N4662
3. Gasket 9N4131
4. Dipstick NAA927C
5. Cover 9N7158C
6. Oil tube 9N567
7. Gasket NAA502A
8. Housing 2N4013A2
   (9N, 2N) 8N4021B (8N)
9. Rocker NCA535A
10. Pin 8N486A
11. Stud 376425S36
12. Oil trough 9N4150
13. Block 8N4194
14. Pin 9N7221B
15. Pin 8N804
16. Clevis 8N802C
17. Drain plug 353064S

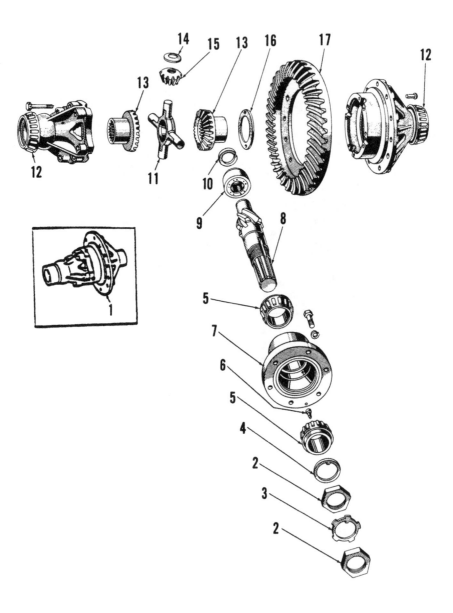

Fig. 196—Differential assembly with ring gear and pinion for 9N, 2N and 8N. Pinion 79-4609 is used with ring gear 79-4210A.

1. Case assy. 9N4204	10. Ring BB4627
2. Nut B4634	11. Spider 9N4211
3. Washer B4636A	12. Bearing BB4221B
4. Washer BB4667	13. Gear B2NN4236A
5. Bearing BB4621B	14. Washer 7RT4230
6. Pin TEAN4256A	15. Gear 7RT4215
7. Retainer BB4614B	16. Washer 79-4228F
8. Pinion 79-4609	17. Ring gear 79-4210A
or 8N4609B	or 8N4210A
9. Bearing BB4625B	

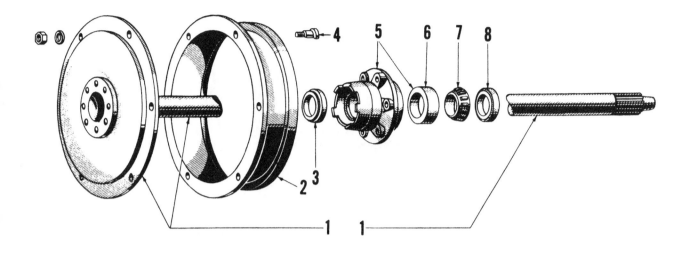

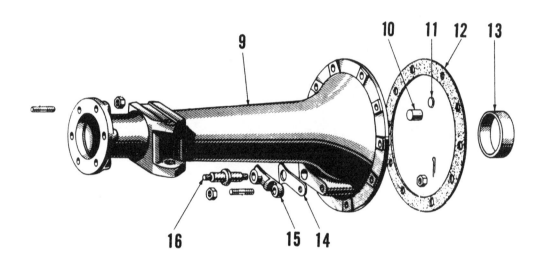

*Fig. 197—Rear axle shaft, housing
and related parts for 9N and 2N.*

1. Rear axle 9N4234B
2. Brake drum 9N1126
3. Seal D6NN4251A
4. Bolt 9N1117
5. Retainer 9N4124
6. Cup 9N4222
7. Cone 2N4221
8. Collar 9N4132

9. Housing 2N4013A2
10. Bushing 78-2477
11. Plug 1-1/4
12. Gasket 2N4035
13. Cup BB4222
14. Gasket 9N523A
15. Plate 9N562B
16. Shaft 9N563

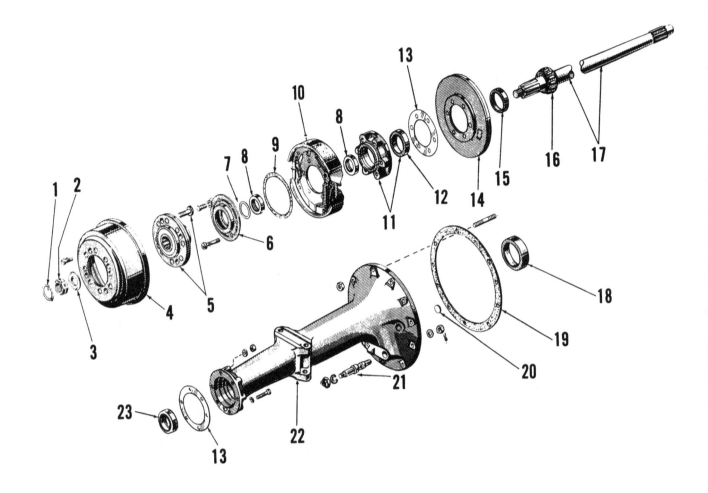

*Fig. 198—Rear axle shaft, housing
and related parts for 8N.*

1. Ring 8N4187
2. Nut C8PN4179A
3. Washer 8N4293
4. Brake drum 8N1126
5. Hub 8N1171
6. Retainer A8NN4248A
7. Gasket 8N4290
8. Seal D5NN4115A
9. Gasket 8N4225
10. Brake shoes 8N2200B
11. Retainer 8N4124B
12. Cup 9N4222

13. Shim 8N4229
14. Shield 8N2255
15. Gasket 8N4284
16. Bearing 2N4221
17. Axle shaft 8N4235C
18. Cup BB4222
19. Gasket 2N4035
20. Plug 1-1/4
21. Shaft NCA563C
22. Housing 8N4010B
     (RH), 8N4011B (LH)
23. Seal 8N4233A1

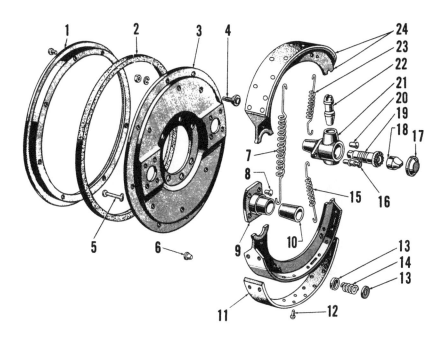

Fig. 199—Brakes and related parts for 9N and 2N.

1. Ring 9N2252
2. Seal 9N2217
3. Plate 9N2213B
4. Equalizer 9N2030
5. Pin 9N2069
6. Plug 3/8
7. Spring 9N2035
8. Rivet 1/4 x 7/8
9. Bracket 9N2304
10. Bushing 9N2227A
11. Brake lining 9N2219A
12. Rivet 351978S
13. Cup 6H2066
14. Spring 9N2068
15. Spring 9N2036
16. Stud 81T2104
17. Cap 81T2043
18. Wedge 9N2041
19. Stem 81T2108
20. Rivet 1/4 x 7/8
21. Bracket 9N2040
22. Link 81T2042
23. Spring 9N2034
24. Brake shoe 9N2219A

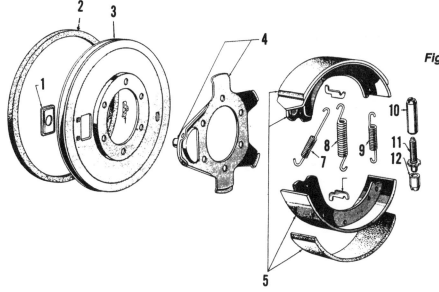

Fig. 200—Brakes and related parts for 8N.

1. Cover 8N2224
2. Dust seal 8N2217
3. Shield 8N2255
4. Plate 8N2212B
5. Brake shoes 8N2200B
6. Anchor 8N2027B
7. Spring 8N2299
8. Spring 8N2297B
9. Spring 8N2049
10. Nut 8N2223
11. Adjusting screw H2041
12. Socket 8N2048

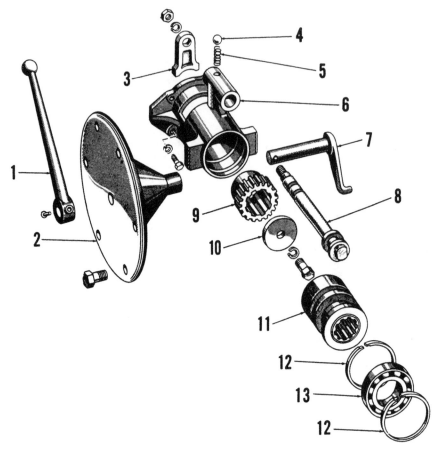

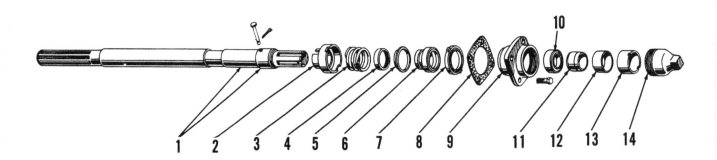

Fig. 201—PTO shifter lever, fork and related parts.

1. Lever 9N723
2. Plate 8N721A
3. Stop NAA722A
4. Ball 353075S
5. Spring 79-7234
6. Support 8N770
7. Fork 9N720
8. Rail 9N719
9. Hub 9N716
10. Washer 351402S
11. Sleeve 2N717
12. Ring 9N754
13. Bearing 9N715C

Fig. 202—Power take-off assembly for 1939 to 1943 tractors.

1. PTO shaft 9N710B
2. Case 9N728B
3. Spring 9N729
4. Ring 9N731
5. Guide 9N730
6. Seal 9N727
7. Washer 9N724
8. Gasket C5NN747A
9. Retainer 9N733B
10. Bearing 9N715C
11. Sleeve 9N735B
12. Spacer 9N714
13. Retainer 9N734
14. Cap 310088

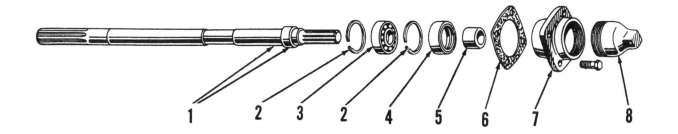

Fig. 203—Power take-off assembly for 1943 to 1952 tractors.

1. PTO shaft 9N710B
2. Ring 9N754
3. Bearing 9N715C
4. Seal D8NN703AA
5. Sleeve 9N735A
6. Gasket C5NN747A
7. Retainer 9N733B
8. Cap 31088

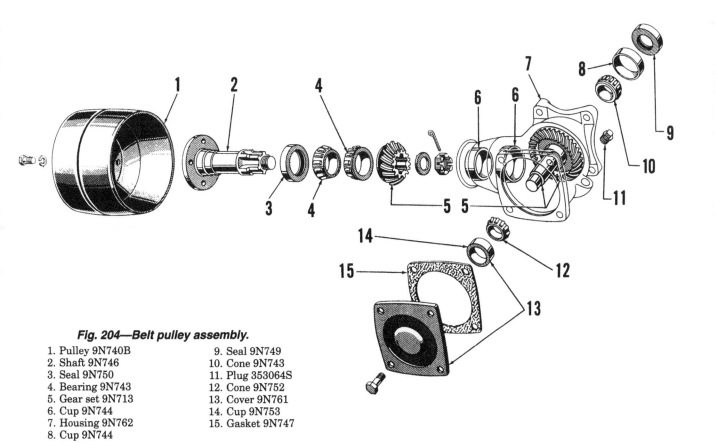

Fig. 204—Belt pulley assembly.

1. Pulley 9N740B	9. Seal 9N749
2. Shaft 9N746	10. Cone 9N743
3. Seal 9N750	11. Plug 353064S
4. Bearing 9N743	12. Cone 9N752
5. Gear set 9N713	13. Cover 9N761
6. Cup 9N744	14. Cup 9N753
7. Housing 9N762	15. Gasket 9N747
8. Cup 9N744	

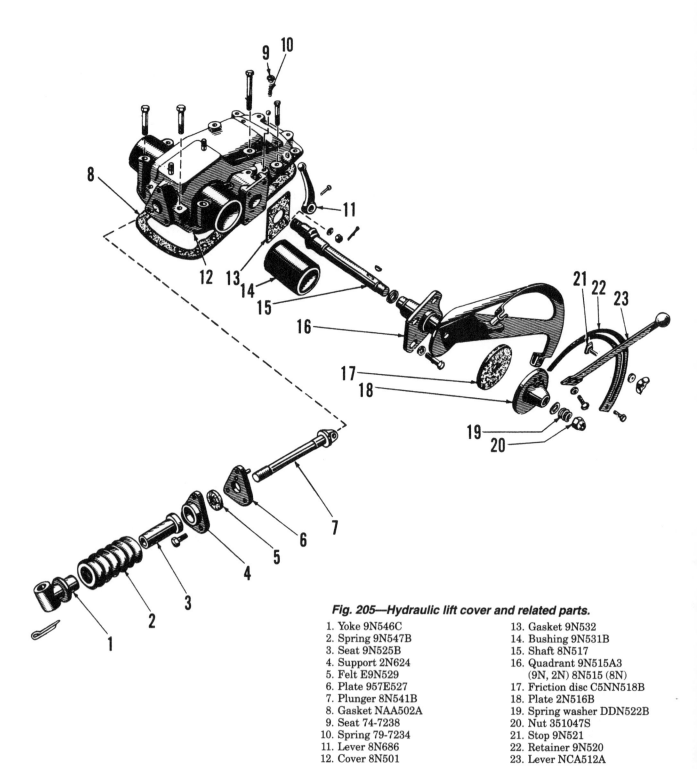

Fig. 205—Hydraulic lift cover and related parts.

1. Yoke 9N546C
2. Spring 9N547B
3. Seat 9N525B
4. Support 2N624
5. Felt E9N529
6. Plate 957E527
7. Plunger 8N541B
8. Gasket NAA502A
9. Seat 74-7238
10. Spring 79-7234
11. Lever 8N686
12. Cover 8N501
13. Gasket 9N532
14. Bushing 9N531B
15. Shaft 8N517
16. Quadrant 9N515A3
    (9N, 2N) 8N515 (8N)
17. Friction disc C5NN518B
18. Plate 2N516B
19. Spring washer DDN522B
20. Nut 351047S
21. Stop 9N521
22. Retainer 9N520
23. Lever NCA512A

Fig. 206—Hydraulic lift shaft and related parts.

1. Cylinder 9N510D
2. Gasket 9N553
3. Ring NAA533A
4. Piston 9N530A
5. Rod 957E526
6. Lever assy. 9N504B (9N, 2N)
7. Fork 9N540 (9N, 2N)
8. Bushing 9N531C
9. Arm D4NN545A
10. Shaft NCA544B
11. Arm 9N543B
12. Plate 9N550
13. Washer 9N551
14. Link 8N540A2
15. Spring 8N539B
16. Lever 8N504 (8N)
17. Nut 34443S
18. Washer 44722S
19. Swivel 8N692
20. Bushing 8N691
21. Cam 8N685
22. Rod 8N683
23. Spring 8N684
24. Arm 8N680
25. Link 8N690
26. Arm 8N693
27. Nut 34420S7
28. Plate 8N689

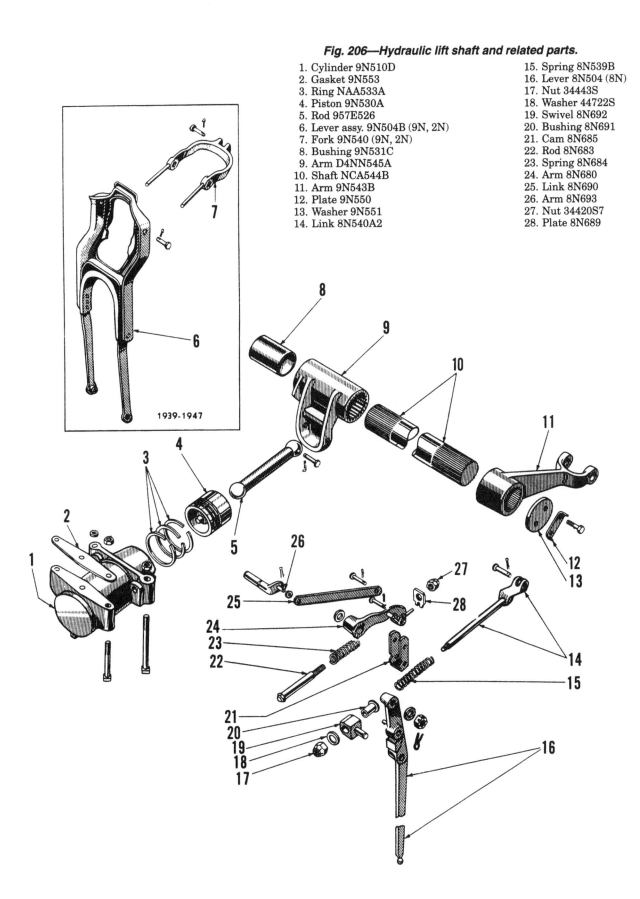

1939-1947

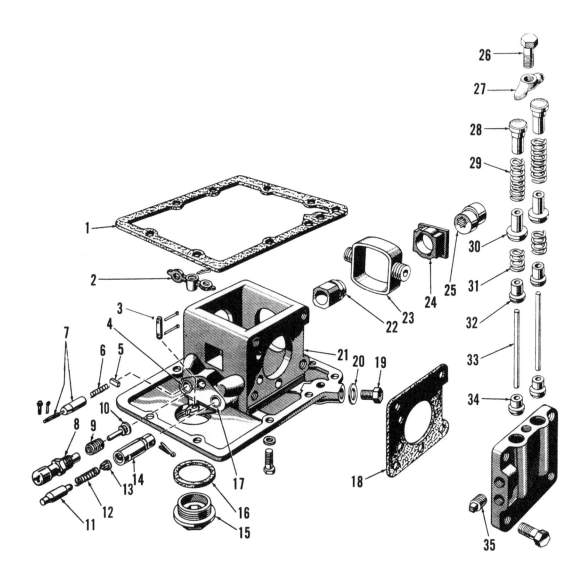

Fig. 207—Hydraulic pump assembly.

1. Gasket 9N611
2. Lever 8N643
3. Pin 357233S
4. Bushing 8N625B
5. Pin
6. Spring 9N632
7. Control valve 8N640
8. Safety valve 8N638
9. Spring 9N645
10. Check valve 9N635
11. Intake valve 8N698
12. Spring 8N697
13. Plug 74211S
14. Bushing .010
　　O/S 8N695BRP
15. Plug 9N7010
16. Gasket 8N7011
17. Bushing 8N695C
18. Gasket 9N613

19. Plug 355650S
20. Gasket C20Z6734A
21. Pump base 9N605A
　　(9N, 2N) 8N605B (8N)
22. Bushing 9N649B
23. Piston 9N615
24. Cam block 9N617B
25. Cam 9N618A
26. Bolt 359264S8
27. Clamp 9N609
28. Plug 9N636
29. Spring 9N646
30. Outlet valve 9N628
31. Spring C7NN647A
32. Inlet valve 9N629
33. Guide 9N626
34. Socket 9N627
35. Plug 87837S36

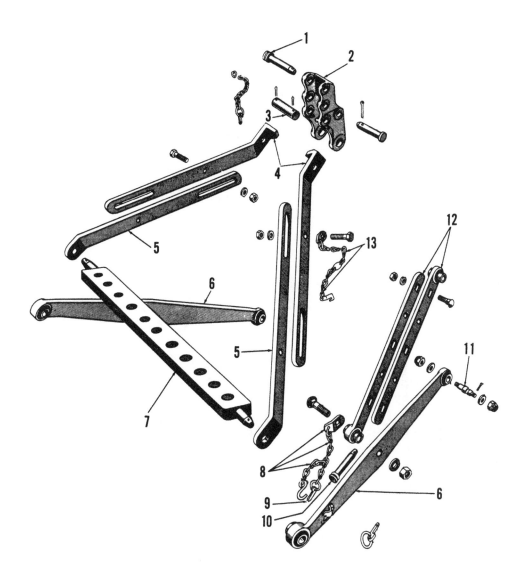

Fig. 208—Hydraulic lift links and related parts.

1. Pin 957E560
2. Rocker NCA535A
3. Pin 8N486B
4. Stay rod C5NNC831A
5. Stay rod C5NNC832A
6. Link 9N555B
7. Drawbar F251006
8. Chain 9N588
9. Pin 2N608
10. Pin 957E560
11. Shaft NCA563C
12. Link 2N666B
13. Chain NCA5370A

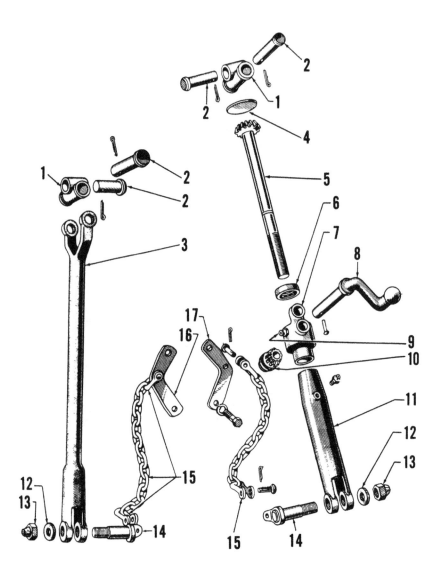

Fig. 209—Leveling gear box assembly and related parts.

1. Knuckle 957E566
2. Pin 9N595
3. Rod 9N564B
4. Plug 1-3/4
5. Gear NCA579B
6. Bearing 957E554B
7. Gear box C5NN565C
8. Lever 9N594
9. Lube fitting 87907S8
10. Gear B9NN568A
11. Fork EC0N585A
12. Washer
13. Nut
14. Eyebolt EC0N660A
15. Chain CBPN598A
16. Anchor 957E597
17. Anchor 957E596

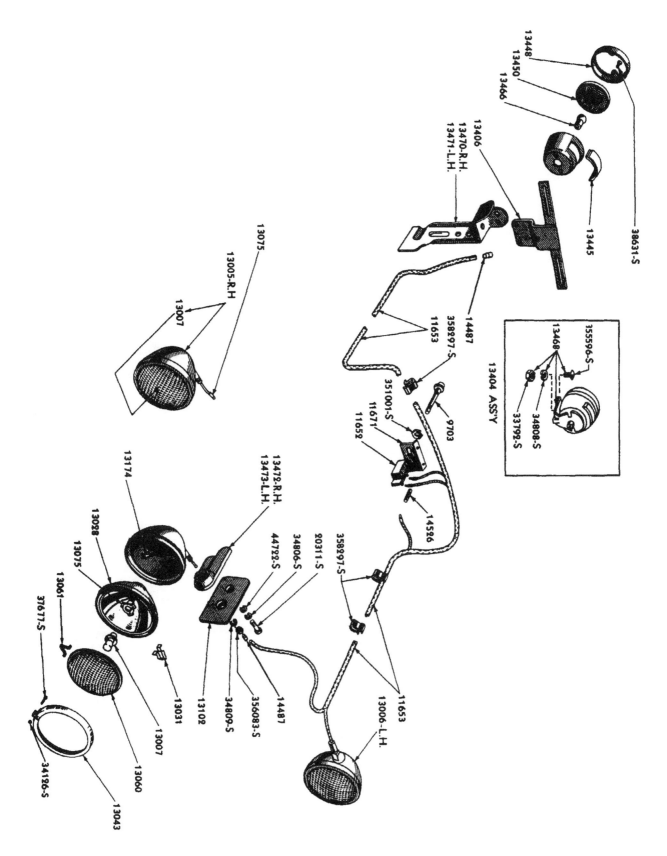

Figure 210—Lighting kit for 1939-1949 models.

147

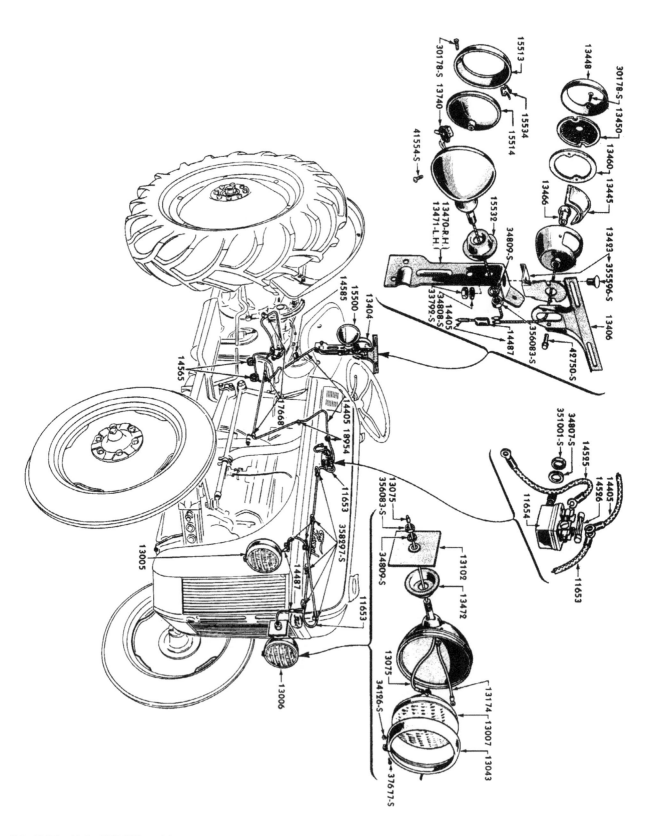

Figure 211—Lighting kit for 1949-1952 models.

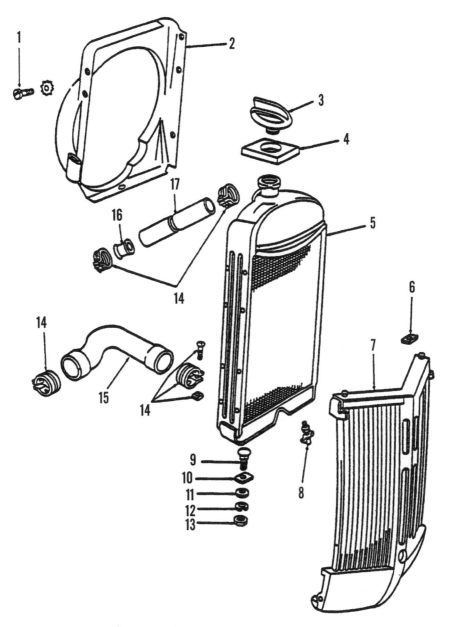

Fig. 212—Radiator and related parts.

1. Pan head screw 8-32 x 3/8
2. Fan shroud 2N-8146
3. Radiator cap 9N-8100
4. Filler neck pad 2N-8032
5. Radiator 8N-8005
6. Bumper (8N only)
7. Grill assy. 8N-8204
8. Drain cock C5NN-8115-B
9. Carriage bolt 7/16—14 x 1 351665-S36
10. Insulator C5NN-8125-A
11. Washer 7/16" (R.H. Side only)
12. Lock washer 7/16"
13. Nut 7/16 x 14
14. Hose clamps 107325
15. Radiator lower hose 312588 (Before SN263844)
15. Radiator lower hose 8N-8286 (After SN 263843)
16. Thermostat B2NN-8575-A
17. Radiator upper hose 2N-8260

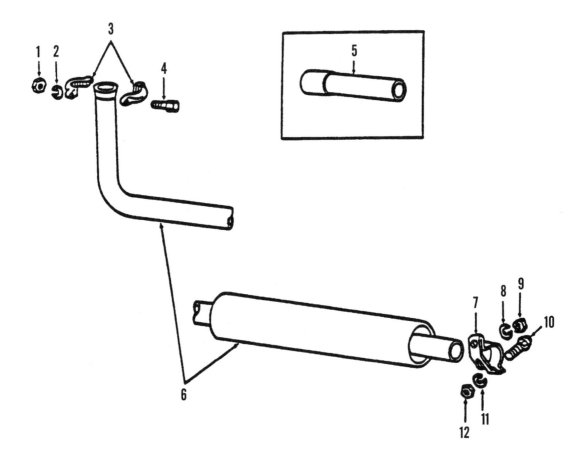

Fig. 213—Muffler and related parts.

1. Brass nut 5/16 x 18
2. Lock washer
3. Muffler clamp CAPN-5200-A
4. Bolt 5/16 x 16 x 1-5/8
5. Muffler outlet pipe 8N-5255
6. Muffler assy. 9N-5230
7. Muffler bracket 9N-6260
8. Lock washer
9. Nut 7/16 x 20
10. Carriage bolt 5/16 x 18 x 1-1/8
11. Lock washer
12. Nut 7/16 x 20

NOTES